Biscuit, cookie and cracker manufacturing manuals

Manual 3
Biscuit dough piece forming

Please personalise your copy with your name below

....

....

Biscuit, cookie and cracker manufacturing manuals

The other titles in this series are:

Manual 1
Ingredients
Types • Handling • Uses

Manual 2
Biscuit doughs
Types • Mixing • Conditioning • Handling • Troubleshooting tips

Manual 4
Baking and cooling of biscuits
What happens in a baking oven • Types of oven • Post-oven processing • Cooling • Handling • Troubleshooting tips

Manual 5
Secondary processing in biscuit manufacturing
Chocolate enrobing • Moulding • Sandwich creaming • Icing • Application of jam • Marshmallow • Caramel • Troubleshooting tips

Manual 6
Biscuit packaging and storage
Packaging materials • Wrapping operations • Biscuit storage • Troubleshooting tips

Biscuit, cookie and cracker manufacturing manuals

Manual 3

Biscuit dough piece forming

Sheeting
Gauging
Cutting
Laminating
Rotary moulding
Extruding
Wire cutting
Depositing
Troubleshooting tips

DUNCAN MANLEY

WOODHEAD PUBLISHING LIMITED
Cambridge England

Published by Woodhead Publishing Limited,
Abington Hall, Abington, Cambridge CB1 6AH, England

First published 1998

British Library Cataloguing in Publication Data
A catalogue record for this book is available from the British Library.

ISBN 978-1-85573-294-0 (print)
ISBN 978-1-85573-622-1 (online)
ISSN 2042-8049 Woodhead Publishing Series in Food Science, Technology and Nutrition (print)
ISSN 2042-8057 Woodhead Publishing Series in Food Science, Technology and Nutrition (online)

Contents

Preface

My text *Technology of Biscuits, Crackers and Cookies*, which was first published in 1983, with an enlarged new edition in 1991, has proved very successful and been welcomed by biscuit manufacturers worldwide. Why, then, consider producing separate manuals in the same field?

The idea started, I suppose, when my partner, Pam Chance, pointed out that, as a standard reference work, my book was both too detailed and expensive for the average plant operative to use in the course of his or her work.

Over the years, I have worked as a consultant in very many factories in many countries. Not all the operatives speak English, but those that do have explained that they often do not know the biscuit technology with which they are involved nor have a convenient source of information. They and their managers have particularly expressed the need for aids to troubleshooting.

Thus this manual was born. It is one of a sequence that covers the various parts of the biscuit-making and packaging process. It builds on *Technology of Biscuits, Crackers and Cookies*, but includes much new material. I have tried to give particular emphasis to process mechanisms and fault solving. I am sure that managers, trainers and operatives will find it useful both in training and as a reference source.

I hope that all who read and use it will find it as useful as I would like it to be. If you have any comments or contributions, I should be pleased to hear from you.

Duncan J R Manley
1998

1 Introduction

This manual is one of a series dealing with materials and manufacturing procedures for biscuits.

It describes, in general terms, what is involved in forming dough pieces from mixed dough. The dough pieces are then baked. The process of making doughs is dealt with in Manual 2 *Biscuit doughs* and the baking of the dough pieces is covered in Manual 4 *Baking and cooling of biscuits.*

If you are a member of a manufacturing team you should know how to do your tasks and the **reasons** for doing things in a specified way. You should also understand the possible implications of not doing a task correctly or not communicating difficulties promptly. The processes involved in forming dough pieces and the critical aspects to watch for during the various stages of these processes are described. It is essential for the efficient running of the factory and the production of biscuits of correct size, weight and texture that the dough pieces are formed correctly. Errors in dough piece forming may not be discovered until after they are baked and thus at least one complete oven full of biscuits may have to be scrapped.

From time to time problems will arise but a competent operator will be able to recognise and correct faults quickly. Sections are included in this manual which describe common faults and problems and should therefore aid in troubleshooting and problem solving.

If you work in a food factory you must accept some responsibilities. These, to a greater or lesser extent, will include:

1 Responsibility for the quality of the company's products if you are a member of a production team. By critical observation and knowing what to look for and expect, you could prevent a fault occurring in manufacturing.
2 Responsibility for the safety of consumers who will eat the

products you help to make and pack. The procedures and precautions you should observe are described.

3 Responsibility for the machines and equipment with which you are required to work. The procedures and precautions you should observe are described.

1.1 Vocational qualifications

The approach to training in industry is changing. Instead of a reliance on knowledge which has typically been assessed by set piece examinations, there is now a focus on competence which is assessed in the workplace. This means that a worker must not only know what he or she should be doing but also has to demonstrate that he or she can do it!

Typically, set piece examinations require the examinee to achieve a 'pass' mark which may be as low as 40%. This accepts the fact that by no means all of the subject matter is known well enough for the examinee to pass the examination. Under the competence system, to qualify, the worker must satisfy the examiner, usually known as the assessor, in all matters. The worker must demonstrate knowledge, ability and communication skills as required for the level of the qualification being assessed. These qualifications are known as vocational qualifications (VQs).

There should be VQs for all employees from the newest and youngest to the most senior. Through training, workers may progress to higher and higher levels. In the developing British system there is a framework of five levels which can be described as:

Level 1 – Competence in the performance of a range of varied work activities, most of which may be routine and predictable.

Level 2 – Competence in a significant range of varied work activities, performed in a variety of contexts. Some of the activities are complex or non-routine, and there is some individual responsibility or autonomy. Collaboration with others, perhaps through membership of a work group or team, may often be a requirement.

Level 3 – Competence in a broad range of varied work activities performed in a wide variety of contexts most of which are complex and non-routine. Often considerable responsibility and autonomy, and control or guidance of others is required.

Level 4 – Competence in a broad range of complex, technical or professional work activities performed in a wide variety of contexts and with a substantial degree of personal responsibility and autonomy. Responsibility for others is often present.

Level 5 – Competence which involves the application of a significant range of fundamental principles and complex techniques across a wide and often unpredictable variety of contexts. Substantial personal autonomy and often significant responsibility for the work of others and for the allocation of substantial resources feature strongly, as do accountabilities for analysis and diagnosis, design, planning, execution and evaluation.

It should be possible to categorise all jobs within a company in one of these five levels. To achieve accreditation at any level it is necessary to satisfy the assessor in a defined number of **units** (each of which has a number of **elements**). There are some mandatory (core) units and some optional units, a defined number of which must be selected, for each level. This reflects the fact that workers at a given level may have jobs that are biased towards production or production support and may be more technical or supervisory. In many respects there will be overlaps between levels and the greatest difficulty tends to arise between levels 3 and 4. The flavour of level 3 should be 'optimise, prioritise and improve' and of level 4 'plan, initiate, develop and manage'.

Thus, it can be seen that as the level of competence progresses there is a change from understanding, to seeking ways of improving and helping the business to become more efficient.

For all jobs and tasks there is a need to define What, Why and How? In biscuit making there are technical aspects which include, what are the ingredients, products, processes and machines, why particular ingredients, packaging materials and processes are needed for different products and how machinery is controlled and maintained. Technical knowledge and skills are needed for problem solving, and to ensure good hygiene and safety in the workplace. Communication skills are also needed which include reporting to and supervising others.

Competence cannot be achieved only from a book but reading and thinking are an aid to learning about ingredients, processes and machinery and understanding what variations may occur. Thus, a book can provide essential underpinning knowledge and is a source of reference when something new or unusual happens.

Using this manual will help you to become a competent employee involved in biscuit manufacturing. The underpinning knowledge that is contained is relevant particularly to most of the technical aspects of levels 2 and 3, as defined above, related to the handling of dough in order to form pieces that will be baked into biscuits.

2 Background to the biscuit industry

2.1 What are biscuits?

Biscuits are small baked products made principally from flour, sugar and fat. They typically have a moisture content of less than 4% and when packaged in moistureproof containers have a long shelf life, perhaps six months or more. The appeal to consumers is determined by the appearance and eating qualities. For example, consumers do not like broken biscuits nor ones that have been over or under baked.

Biscuits are made in many shapes and sizes and after baking they may be coated with chocolate, sandwiched with a fat-based filling or have other pleasantly flavoured additions.

2.2 How are biscuits made?

Biscuits are a traditional type of flour confectionery which were, and can still be, made and baked in a domestic kitchen. Now they are made mostly in factories on large production plants. These plants are large and complex and involve considerable mechanical sophistication. Forming, baking and packaging are largely continuous operations but metering ingredients and dough mixing are typically done in batches.

There is a high degree of mechanisation in the biscuit industry but at present there are very few completely automatic production plants. This means that there is a high degree of dependence on the operators to start and control production plants. It is essential that operators are skilled in the tasks they have to do and this involves responsibility for product quality. As part of their training they must know about the ingredients and their roles in making biscuits. They must be aware of the potential ingredient quality variations and the significance of these.

There are basically two types of biscuit dough, hard and soft. The difference is determined by the amount of water required to make a dough which has satisfactory handling quality for making dough pieces for baking.

Hard dough has high water and relatively low fat (and sugar) contents and the dough is tough and extensible (it can be pulled out without immediately breaking), like tight bread dough. The biscuits are either crackers or in a group known as semi-sweet or hard sweet.

Soft doughs contain much less water and relatively high levels of fat and sugar. The dough is short, (breaks when it is pulled out) which means that it exhibits very low extensible character. It may be so soft that it is pourable. The biscuits are of the soft eating types which are often referred to as 'cookies'. There are a great number of biscuit types made from soft doughs and a wide variety of ingredients may be used.

The machinery used to make biscuits is designed to suit the type of dough needed and to develop the structure and shape of the individual biscuits.

Secondary processing, which is done after the biscuit has been baked, and packaging biscuits are specific to the product concerned. There is normally a limited range of biscuit types that can be made by a given set of plant machinery.

Many biscuit production plants bake at the rate of 1000–2000 kg per hour and higher rates are not unusual. Given this and the sophistication of the production line it is most economical to make only one biscuit type for a whole day or at least an eight hour shift. Start-ups and changeovers are relatively inefficient.

2.3 How a factory is arranged

Typically the factory is long and, for the most part, normally on only one floor. The reason for the length is principally due to the oven.

Tunnel ovens have baking bands that are usually between 800–1400 mm (31–55 in) in width. The length of the oven determines the output capacity of the plant. Ovens have been made up to 150 m in length but 60 m (about 200 feet) is probably the average length.

Ideally, and normally, the ingredients are stored and handled at one end of the factory. Next to the ingredients store is the mixing area and next to that are the continuous production plants. The baking plants feed cooling conveyors, which are often multitiered to save space, and the baked and cooled biscuits are then packed using high speed machines.

In some factories secondary processes are involved after baking. It is also possible that only semi-automatic packaging is used which requires manual feeding of the wrapping machines. In these cases biscuits may be taken from the baking line and placed temporarily in boxes or stored in other ways. These activities are typically labour intensive.

2.4 What your company requires from the factory

Your company exists to make a profit! The means of earning this profit is by making and selling biscuits (and possibly other products).

The products that are made are designed to meet current market needs and to this end they have specifications in terms of pack size, biscuit eating qualities and appearance and ingredient types and quantities. These specifications define limits and it is the task of the production department to ensure that only biscuits which meet the specifications are packed and sold. All substandard product must be disposed of through other routes and will represent a financial loss to the company.

The production cost of a product is a combination of ingredient and packaging material costs, labour (which involves the production time), fuel for the machines and baking, and overheads which include management, maintenance and other support services. A significant cost is the labour associated with the production time. The efficiency of production, which is measured by the quantity of saleable product produced in a specified time, is an important aspect of the product cost.

The duty of the Production Department to the company is to produce a minimum of scrap product and have a minimum of production downtime. Both of these requirements are influenced strongly by the skills and performance of the plant operators.

2.5 Dough piece forming areas

Dough piece forming is usually achieved using a series of machines placed in line immediately in front of a tunnel oven. The machinery used to roll, laminate, cut or mould dough into pieces plus the oven and any cooling conveyors is collectively known as the biscuit plant. In bakeries where there is more than one plant it is normal to arrange that they run parallel to one another and therefore all the dough piece forming sections are located together at one end of the bakery. Plants are designed to be controlled from either the right or left side and commonly adjacent plants are in an opposite control configuration to allow one operator to control two plants.

Dough is delivered to the head of each plant either directly from the mixing machines or in tubs or on conveyors or from holding areas where the dough will have been in tubs. Often the dough is transferred to the dough piece forming machinery on the floor above so that it falls down through a hopper or dough feeder. In older plants transfer of dough to the plant is by hand from tubs. This is hard work and potentially more unhygienic. Great care must be taken that nothing falls into the dough because if anything is lost in the dough it may appear as an unpleasant or dangerous inclusion in the baked biscuits.

It is inevitable that from time to time some pieces of dough will fall onto the floor. This dough is potentially dirty and should never be returned to the main mass of dough in the plant. Dough on the floor will also become spread about on shoes and may cause the floor to become slippery. It is thus important that trays be placed to reduce the chance of dough getting onto the floor and to assist in its removal and safe disposal. How machines are fixed to the floor and the location of catch trays have an important bearing on the ease with which bakeries can be cleaned and how infestation by insects and other animals can be controlled.

2.6 Your contribution when machining doughs

The most important contribution you can make to the efficiency of the factory when machining doughs is to ensure the following:

- Understand the critical points to watch for at each stage where the dough passes through the machines and know how to make adjustments to keep the plants running correctly and continuously.
- Know how to set the machines for a particular product at the beginning of production so that the production starts smoothly and on time.
- Know how to alter the dough piece weight, how to take measurements of dough piece weights and how and when to respond to requests for adjustments to either weight or other aspects of shape. These requests will come from the operator at the exit of the oven, usually known as the baker, because it is there that the final product will be seen and can be assessed for the first time.

- Try to ensure that a minimum of dough falls from the plant as this will be either wasted or will require special handling. You should remember that the weight of the dough pieces affects the efficiency of a plant more than any other single factor. Biscuits of incorrect weight will result in packs that are either too light (a legal offence) or too heavy (representing an extra cost for the company). Incorrect weight will probably mean that the biscuits are too thick or too thin and may have the wrong moisture content after baking. All these factors affect the eating quality and probably the ability to pack the biscuits efficiently.
- Watch for changes in the appearance of the dough and dough pieces as this may reflect the need for machine adjustment or that there is a difference in the mixed dough due to a fault or abnormal standing time. Such changes will almost certainly cause changes in the quality of the baked biscuits.
- Look out for machinery that requires maintenance for reasons of oil leakage, malfunction or breakage and report the problem without delay to your supervisor.
- Avoid straining yourself; do not try to move heavy weights without help or using the appropriate machinery.

In a well managed factory you will probably be required to record various aspects of what you have done. For example, the times that the plant was started and stopped and, if during a normally continuous run, reasons for the stoppage. In this way there will be no misunderstanding about what was done and when and may help in identifying where engineering maintenance is required.

It is not unusual to find faults or to have queries. You must communicate.

It is essential that if you are in doubt you should not hesitate to ask, even if you know that you should know the answer!

3 Hygiene and safety aspects

The regulations relating to food production are continually being tightened with the aim of improving the safety of food products and the safety of people working in food factories.

If you work in a food factory you must be acquainted with the potential dangers and constantly endeavour to prevent the food becoming contaminated with noxious substances. You must also make sure that your actions do not put yourself or your colleagues at risk of injury.

3.1 Safety of food products

Biscuits will be unfit or unpleasant to eat if they are contaminated in the course of their manufacture and packaging. Contaminated means that unwanted material becomes included in or on them. Some forms of contamination may be positively dangerous to the health of those who eat the biscuits.

It is therefore important that the problem of contamination is considered because it is the basis of food hygiene which is the responsibility of all who work with food.

It is not possible to list all the possible hazards to hygiene that may be encountered in a biscuit factory but the following section should help to make you aware of the likely problem areas.

3.2 Sources of contamination

3.2.1 People

- Contamination may come from people via the microorganisms on their hands. Hairs, buttons and pieces of jewellery may fall from their bodies and clothes and articles may fall from pockets.
- The most important requirement for all those who handle, or are likely to handle food, is to observe basic rules of personal hygiene.

Disease is quickly spread if food handlers are negligent about hand washing following visits to toilets. It is very unpleasant to have food contaminated with grease or other dirt from unwashed hands.

- At all food premises good, clean washing facilities must be provided with continuous supplies of hot and cold water, non-scented soap and disposable towels. Cold water with no soap and communal towels are not adequate.
- Hand washing sinks and facilities must be separate from those used to wash equipment.
- **All food handlers must ensure that their hands are washed and clean before handling food and it is particularly important that their hands are washed after each visit to the toilet.**
- Employers must provide clean overalls and hair coverings for all personnel. These are to be worn only in the food factory. No personal food, drink containers, loose money, pins, jewellery (other than plain wedding rings), watches, radios, books, newspapers and smoking tackle should be allowed into the production areas. Hair brushing or combing necessitating removal of head gear should be forbidden in production areas. **In this way the possibility of contamination by loose articles is significantly reduced.**
- Smoking involves the hands becoming contaminated with saliva and the by-products – matches, ash and cigarette ends, are particularly repulsive. **No smoking should ever be allowed in the production areas.**
- Operators who have cuts, abrasions or skin infections, particularly on the hands or arms, should be especially careful. **Bandages or dressings should be of good quality and be, at least partly, brightly coloured and easily detectable should they be lost.**

  In those premises where metal detectors are available for product scanning, it is additionally useful for the bandages to contain metal strips that will be found automatically should a bandage be lost in the product.
- **Food handlers suffering from intestinal complaints** such as diarrhoea or other contagious diseases should be required to keep away from production areas until they recover.
- It is frequently necessary for operators to carry certain small articles with them in the course of their duties. **Articles such as pens, pencils, gauges and various tools should not be**

carried in top pockets in case, while bending over, they should fall into the product or machines. Overalls provided with no top pockets remove this possibility!

- Where gloves are needed either of fabric type (as for chocolate handling) or waterproof, they require regular washing and drying both inside and out. **Gloves should not be used by more than one person and they should be replaced when damaged**.

3.2.2 Emptying containers

- When bags or boxes are opened and emptied there is a great potential for contamination.
- Pieces of string or paper removed in the opening process must be placed in rubbish bins and not on the floor.
- Before inverting a bag, box or other type of container, ensure that it has not collected floor or surface dirt that could fall into an unwanted place.
- Dispose of the empty container in a responsible way so that spillage or dust is avoided as much as possible and it is not a danger to other workers.

3.2.3 Small items of equipment

- In most biscuit factories it is necessary to use bowls, beakers or trays to carry and weigh ingredients or dough. These should be of metal or plastic because glass is particularly dangerous, making splinters or small fragments if broken.
- **Glass containers must never be taken into production areas**. Where ingredients are delivered in glass containers they should be dispensed into non-breakable containers in specially designated rooms away from the production areas.
- Colour coding of containers is better than labels which may fall off. Elastic bands provide a particular hazard due to their tendency to fly off in unexpected directions and become lost.
- All utensils should be stored, full or empty, on special clean stillage so that they are off the floor. This is to ensure that when inverted no floor dirt can fall from them on to product or into a mixer.
- After use all containers should be washed in hot water, with detergent as necessary, and left inverted to dry.

- Cleaning equipment such as cloths, brushes, mops and scrapers should be stored and dried after use on specially provided racks, hooks or rails, off the floor.
- Detergents used for cleaning equipment must be of approved types and stocks must be stored separately away from ingredients or dough containers.
- Office equipment such as elastic bands, paper clips and particularly pins should be forbidden in the factory environment.

3.2.4 Plant machinery

- At the end of each production run all machines should be cleaned immediately so that buildup of dough or other materials does not become hard, or mouldy and an attraction for insects.
- As a basic principle, all food machinery should be mounted off the floor so that the floor can be thoroughly swept or washed at regular intervals.
- Covers for the moving parts of machinery should be properly fixed at all times and kept in good repair.
- All surfaces should be wiped down regularly and washed with warm water and detergent if necessary.
- Fabric conveyors should be checked regularly to watch for frayed edges or seams. If necessary these should be trimmed with a sharp knife or the conveyor replaced.
- If a machine is not to be used for some time it should be covered with a dust sheet.
- Drip trays and other catch containers must be emptied and cleaned regularly, but certainly at the end of each production run.
- Particular care should be taken that mineral lubricating oils and greases do not contaminate food. Leaking motors, gearboxes or bearings should be reported without delay for engineering maintenance.
- Where it is necessary to climb up to high parts of machines or where ladders are needed to get over machines, special walkways with adequate guarding should be provided to prevent floor dirt, carried on footwear, dropping on to dough, products or food surfaces.
- No string should ever be used to attach wires or other articles in production areas and fibrous or loose insulation materials should be covered and fixed securely to prevent disintegration.

- Nowhere in production areas should wood be used. This is easily splintered and pieces find their way into ingredients or dough.
- As machinery is replaced or becomes obsolete it should be completely removed from the production area and stored (preferably in a reasonably clean condition) in a store remote from the factory. In many factories the machinery graveyard is an ideal home for marauding rats, mice and insects. It is ideal from their point of view because it is dry and undisturbed. Food can be taken there and breeding take place in relative comfort! The convenience of such a home should be denied within the production environment.

3.2.5 Buildings and general factory areas

A major source of contamination is from insects, animals and birds. Also dirt or loose particles falling from overhead areas offer potential hazards.

- Flying insects and birds must be excluded from the factory by using screens over ventilation fans and windows which open.
- Open doorways should have plastic strip or air curtains to prevent entry of insects and birds.
- Doors to the outside should fit closely to the floor so that animals cannot enter at night or other times.
- Rodent control systems should be regularly maintained and any bait must be placed only in specially designed and sited containers which are clearly marked. Damaged bait containers should be disposed of immediately and safely.
- Trunking for wiring and other services should be well sealed to reduce the chance of dust accumulation followed by insect infestation.
- High ledges and roof supports where dust can collect should be of sloping construction and be cleaned regularly.
- Good lighting should be maintained in all production areas and plastic screening, where appropriate, should be used to prevent glass falling on to the product if light bulbs or tubes are broken.

It is a requirement that all food workers are aware of these potentials for contamination and that they report to management, without delay, any aspects that do not seem satisfactory.

3.3 Safety of people

Your employer is required to ensure that the areas in which employees work and the machinery they use is safe. However, if employees are negligent in reporting faults or in cleaning operations, etc. it is possible that an otherwise safe situation can become unsafe. You should therefore be aware of where unsafe situations may arise which could affect you or your colleagues.

3.3.1 Floors

Dirty floors which have become wet or greasy are slippery. Clean up as necessary.

3.3.2 Machine guards

Moving parts especially those where a nip is involved must be guarded to prevent hands or clothing becoming trapped. It is particularly dangerous to run a machine with these guards removed. Experience shows that accidents involving machines occur more often to 'experienced' operatives. They become over confident and try to overcome problems by running machines with guards removed.

3.3.3 Electrical connections

Most machinery is driven and controlled by electricity. For safety and other reasons the connections and other electronic components are housed in cabinets or under guards. The danger of electricity cannot be seen, so it is very dangerous to remove guards. Faults in electrical apparatus must be reported to management or responsible engineers.

3.3.4 Strain injuries

Back strain is a common injury experienced by factory workers. It is unpleasant for the person who receives it and a potential cost to the employer due to the need for sickness leave.

Back strain derives from physical effort done incorrectly or carelessly. Think when moving and lifting bags, boxes or pieces of machinery. If they are too heavy get help. If the floor is slippery take extra care. Do not try to lift something too high without help. Do not expect a colleague to help you if he or she is not clear what is expected or is not strong enough to do it.

3.3.5 Dust

Dirt is defined as material in the wrong place! Dust soon becomes dirt. It is unpleasant, may be dangerous to breath and may accumulate and fall into containers bearing food or ingredients. Make sure that your actions cause as little dust as possible and clean up after you.

3.3.6 Building maintenance

When repairs and decoration are being done it is essential that nearby production equipment is covered with sheets so that particles of metal, glass, masonry and paint do not fall into places where they could later be included with dough or product. A magnet provides a useful means of collecting most metal particles.

4 Problem solving

4.1 Introduction

Biscuits are usually baked in vast numbers in continuously running ovens. It is necessary to maintain a constant vigilance to ensure that the biscuits are made with minimum variation. If this is not done, or is done badly, biscuits may be made that are not suitable for consumption or that will not fit into the packaging. From time to time it is necessary to make adjustments and alterations to the process or dough formula to maintain the biscuits within specification. It is the duty of plant operators to make these adjustments. The efficiency of a factory is largely measured by the skill of operators in making these adjustments correctly, accurately and speedily.

In most cases problems are not caused by one factor alone. They are the result of interactions between several factors including ingredients, machine settings and other processing conditions. This means that check lists for both identifying and solving problems are often not simple. Ideally it would be good to have an interactive problem-solving computer program that would interrogate the operator(s) and offer a set of instructions that would lead to the removal of the problem. Not all factories have computers available on the factory floor. Where computers are not available, it is suggested that **reference charts and lists** are kept as a back up to experience and memory of what happened previously. Having observed a problem, the operator can then use the chart or list as a guide to further action. However, understanding why a fault or problem may have arisen is better assessed with some background knowledge. Reference lists to problem solving may be found in each section where a process is described and a reference can be found on each problem solving chart to the section in this manual that provides more information about the parts of the processes involved.

4.2 The process audit record

Much difficulty is encountered at plant start up. A slow start is inefficient and can be costly to the factory. There is therefore a need to set the plant so that the start up is as smooth as possible. Drawing on previous experience is clearly the best approach and to this end a process audit record (see Fig. 1) should be available for each plant for each biscuit that is made. It will give details of machine settings and speeds and the values of process measurements that have to be taken. From time to time the validity of the information on the process audit record will be checked and reviewed because processes can change as a result of time of year and quality of ingredients, etc.

4.3 Control philosophy

When faults and difficulties occur it is necessary to determine their cause and to act as quickly as possible to cure them. The aim in this manual is to enable quick fixes, not detailed research programmes that will take a long time and lead to significant changes in processing methods. It is assumed that production of the products in question has proceeded well for long periods but problems have arisen, which if not attended to, will cause inefficiency or waste.

First, the nature of the fault or problem must be identified. If a pattern in the occurrence of the fault can be found, its cause, and thence its remedy, is more likely to be found. The nature of the patterns of faults and problems will be discussed as appropriate.

If the control philosophy is wrong, the order of making adjustments will not solve the problems that are encountered satisfactorily. Although this manual relates to dough piece forming it will be appreciated that this part of the biscuit making process cannot be viewed in isolation. Therefore, an overall statement of control philosophy follows.

Control implies adherence to product specifications and production targets. The data on the process control chart will show details of the plant and process when, on a single occasion, the whole plant, not just the dough piece forming area, was running well. It will therefore be a useful reference point if difficulties are encountered.

To maintain a plant in control the following principles should be observed:

- Make measurements but only where and when required.

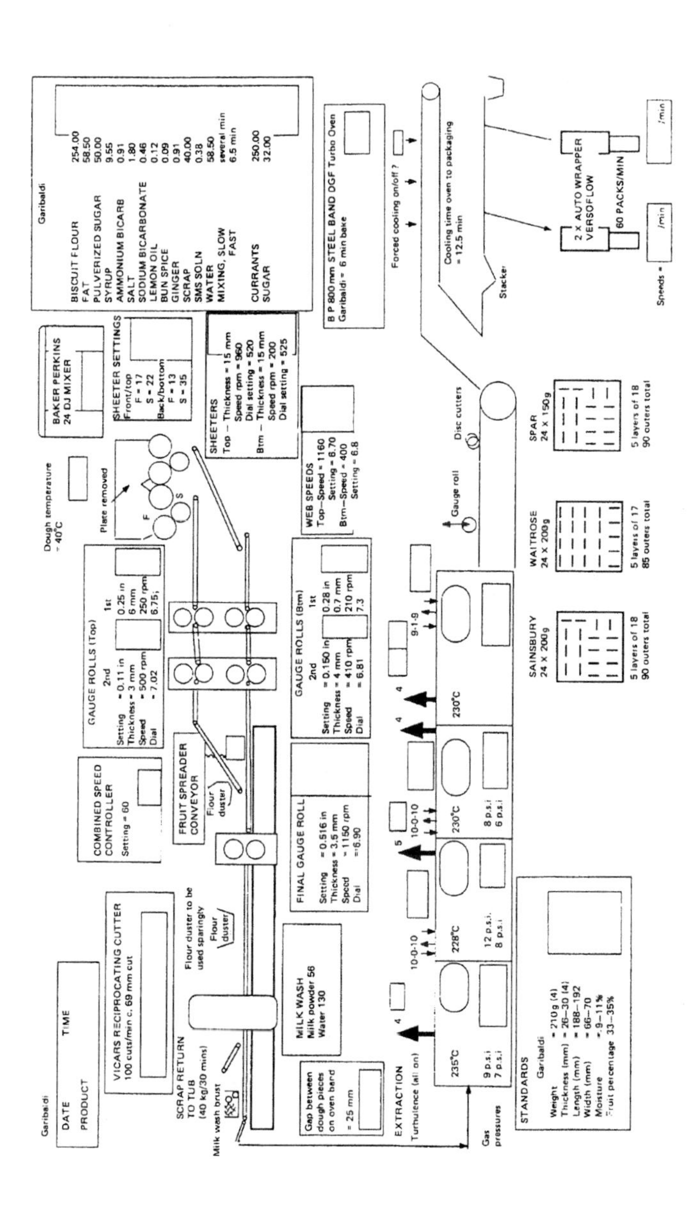

1 Process audit record for production of Garibaldi biscuits

- Relate sampling and measurements to subsequent processing requirements.
- Consider the significance of the results of measurements with respect to those against that can reasonably be achieved with the instruments being used. For example, if the reproducibility of a test is poor then small differences in measurements may not mean that changes are occurring.
- Relate measurements to targets which have specified limits.
- Try to understand what affects the parameter being measured and therefore what type of action can be taken when necessary.
- Concentrate process control surveillance on baked biscuits at the oven exit.
- Biscuit weight is the primary parameter. Changes in weight will probably result in changes in other measurable characteristics like size and colour so concentrate on controlling weight before other characteristics.
- Concentrate process control surveillance in those places where significant variation is known to occur or can be expected.
- Record **all** measurements against time, preferably on a chart which allows easy viewing of changes and trends.

The basic measuring tools applicable to dough piece forming are:

- a balance for weighing dough pieces,
- a calliper gauge for checking thicknesses and dough piece sizes,
- a stop watch for measuring and checking cutter speed,
- a clock telling the time of day.

5 Sheeting, gauging and cutting

5.1 Principles and machinery control systems

Sheeting, gauging and cutting provide the most versatile and commonly used method of the various means of forming pieces for baking from a mass of dough. It is nearly always used for developed or fermented doughs and sometimes for short doughs.

After mixing, the dough may be rested, allowed to ferment or cure, or it may be taken immediately to the hopper of the sheeter. The function of the sheeter is to compact and gauge the mass of dough into a sheet of even thickness and at full width of the plant. There must not be any significant holes and the edges must be smooth and not ragged. Often the sheeter also enables the incorporation of dough returned from the cutter, known as cutter scrap, with fresh, or virgin dough brought from the mixer.

Within the sheeter the dough is compressed and worked to remove air and it is inevitable that some stresses are built up in the gluten structure.

The new sheet of dough passes to one or more sets of gauge roll pairs which reduce the thickness to that required for cutting. Usually the dough is carried between the sheet and each of the gauge rolls on conveyors. Sometimes, having been reduced in thickness, the sheet is folded to form many laminations, before being further gauged to a final desired thickness. For simplicity of description laminating is described separately in Section 6.

Each gauging station adds further stresses to the dough sheet and usually there is insufficient time for these stresses to be relieved naturally before the next stage of processing is reached. The way in which the dough sheet is fed into the gauge roll affects the amount of stress put into the dough. Careful and precise control is needed if variable stresses, which tend to persist all the way to the cutter and therefore may affect the baked biscuit shape, are to be avoided.

The whole series of machines, from sheeter to cutting and finally to panner which puts the dough pieces onto the oven band, is referred to as the 'cutting machine'. As the sheet of dough is made thinner in its progress from the sheeter to the cutter it must become longer. Thus, each gauge roll and the subsequent conveyor runs faster than those before it.

Precise control of the speeds of the various machines and conveyors is essential for smooth running of a cutting machine. They are linked together and the speed ratios are cascaded such that if the speed of one machine is changed all those behind it are changed proportionally. Usually the speed cascade is achieved by electronic control but older machines had PIV (potentially infinitely variable) gearboxes to give similar but not such precise control.

Between the last, or final, gauge roll and the cutter, special provision is usually made to allow relaxation of the dough before the pieces are cut out. During this relaxation the dough shrinks and thus thickens so the thickness at which the sheet is cut, the main factor in determining the dough piece weight, is dependent on both the gap at the final gauge roll and the amount of relaxation allowed. However, the main reason for providing relaxation is to control the shape of the biscuit after baking. A dough sheet that is under tension at the time of cutting will produce dough pieces that show shrinkage in length in the oven and will tend to be thicker at the front and back as a result. If most of the tension is relieved before cutting, the shrinkage will be less severe and the uneven thickness of the biscuits will be less noticeable. By varying the amount of relaxation, the length and shape of the biscuit can be controlled to a certain extent.

Cutting produces not only the outline of the desired shape and size, but also the surface imprint and docker holes. It is necessary to ensure that the dough piece adheres preferentially to the cutting web and not to the cutter. This adherence must not be too severe otherwise difficulty is experienced in transferring the pieces, without distortion, onto the next conveyor or the oven band. Between the cut pieces is a network of unwanted dough known as cutter scrap. This scrap is lifted away and returned either to the sheeter or, less commonly, to the mixer for reincorporation with fresh dough. As the density, toughness and perhaps the fat content and temperature of scrap dough is often different from the fresh dough, it is important that its incorporation with the new dough is as uniform as possible. In some cases it is best to incorporate this scrap preferentially in the top surface of the new sheet and at other times in the bottom surface.

Scrap dough nearly always gives process control problems. Its reincorporation should be planned carefully.

It is possible to sheet, gauge and cut most short doughs as well as extensible cracker and semi-sweet types. The network of cutter scrap from these doughs is a relatively low percentage of the whole dough sheet and it can be lifted and handled after the cutter relatively easily. However, when cutting short dough the cutter scrap often presents problems. During baking, short dough pieces often spread to a larger size and this means that the spacing between the pieces must be larger than for hard dough pieces at the cutter. This results in a higher percentage of cutter scrap that must be handled and reused. Short dough should be 'worked' as little as possible to achieve best quality biscuits so the handling of the cutter scrap when sheeting and cutting of short dough is of fundamental importance to process control. Also, short dough by its very nature is difficult to lift in thin strips. Handling this dough at the point where the cutter scrap is removed from the cutting conveyor, can be very critical and may require special techniques.

Commonly the dough sheet or dough pieces are garnished with sugar, salt, nut fragments, grated cheese, etc., or are given a wash of milk or egg before being baked. Such applications must be made uniformly. If they are made before cutting they should not affect the performance of the cutter or have too much effect on the quality of the cutter scrap as it is reincorporated with fresh dough. A typical biscuit cutting machine is shown in Fig. 2.

5.2 Sheeting and gauging

5.2.1 Dough feeding systems

The dough may be fed to the plant by placing it directly into a hopper above the sheeter. In other cases, dough is fed into a pre-sheeter to allow better control of the feed to the sheeter and to allow incorporation of the cutter scrap to be more uniform. Typically pre-sheeters have two rolls and do not deliver a complete and cohesive sheet of dough. Instead they deliver pieces of dough onto a conveyor which passes to the hopper of the main sheeter. The pre-sheeter and the subsequent conveyor are normally started and stopped to maintain a constant level of dough in the sheeter hopper (see Fig. 3).

The main disadvantage of such a dough feed system is that the dough is exposed to the air for a longer time and may cool or dry.

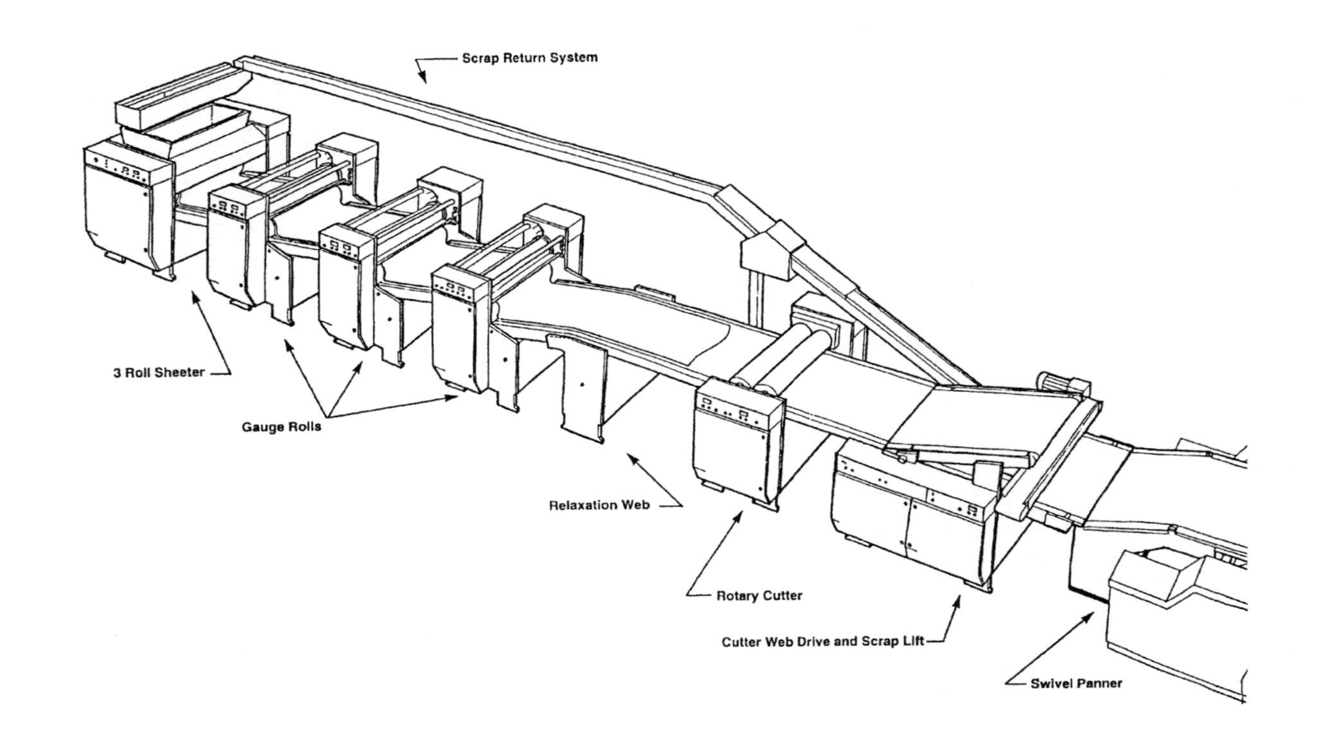

2 A typical biscuit cutting machine

5.2.2 Sheeters

Sheeters are available with either two, three or rarely four rolls. The two roll varieties are usually used as pre-sheeters, that is they meter the dough from a hopper as a rough or incomplete sheet to other machinery such as a rotary moulder or another sheeter at the head of the forming machine. The performance of pre-sheeters is usually not critical as they are not designed to produce a perfect sheet of dough.

Sheeters heading the 'cutting machine' are nearly always of the three roll variety because, as Fig. 4 and 5 show, the configuration of the rolls is designed to compress and gauge the dough into an even full width sheet. The two top rolls are known as forcing rolls and one side of these rolls plus the lower third roll constitutes the gauging facility. In order to draw the dough into the sheeter at least one of the forcing rolls has a rough surface in the form of fluting or grooving. If both forcing rolls have grooves a pattern will be made on one surface of the emerging dough sheet. Generally this is not desirable if the pattern is on the top of the sheet as the effect may persist at the cutter and then affect the appearance of the biscuit surface. The gauging roll always has a smooth surface.

Figures 4 and 5 show that there are front and back discharge types of three roll sheeters. The front discharge variety is preferred for an extensible dough but back discharge is required where the dough is weak and short and needs to be well supported and not bent as the sheet is transferred to the following conveyor.

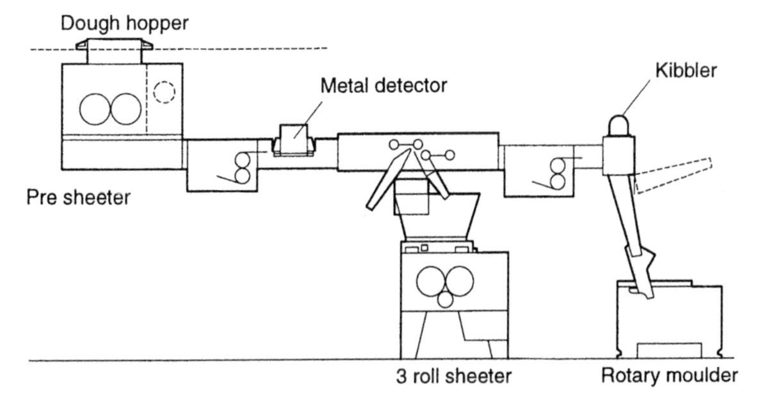

3 Pre-sheeter and dough feed arrangement.

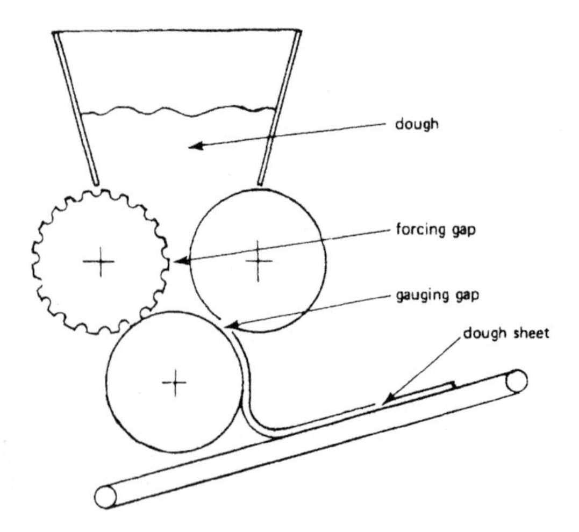

4 Front discharge sheeter.

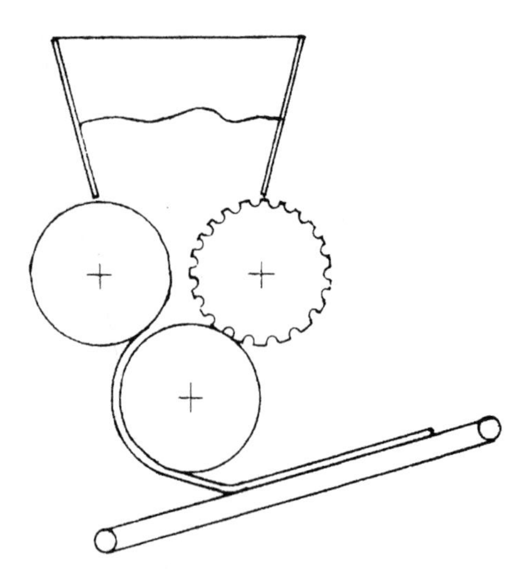

5 Back discharge sheeter.

Four roll sheeters are essentially a three roll sheeter with an extra roll below the gauging roll. When the unit is being used as a sheeter the lowest roll is not used and is merely a support for the conveyor. The two lowest rolls act as a gauging pair when dough is fed from behind the sheeter, for example, from a laminator.

Two roll sheeters do not have the forcing or compression facility and consequently are apt to give sheets which are holey or have ragged edges.

The shape of the hopper which feeds the sheeter is more important than is generally recognised. Figures 6 and 7 show two basic hopper shapes.

A typical hopper, with slopes to increase the dough capacity, often results in the dough becoming 'bridged' over the nip into the forcing gap. This inhibits and often completely stops the flow of the dough into the sheeter. The hopper with vertical sides over the top dead centre of the forcing rolls is much less likely to result in bridging. One or two slowly revolving 'breaker' shafts are commonly provided at the base of the hopper to prevent the dough from bridging. Generally these work well but care must be taken when short dough is being handled as these breaker shafts can result in too much 'working' of the dough, resulting in toughening and reduction in biscuit quality.

It should be noted that the greater the mass of dough in a hopper, the more the pressure at the forcing gap and, therefore, the more

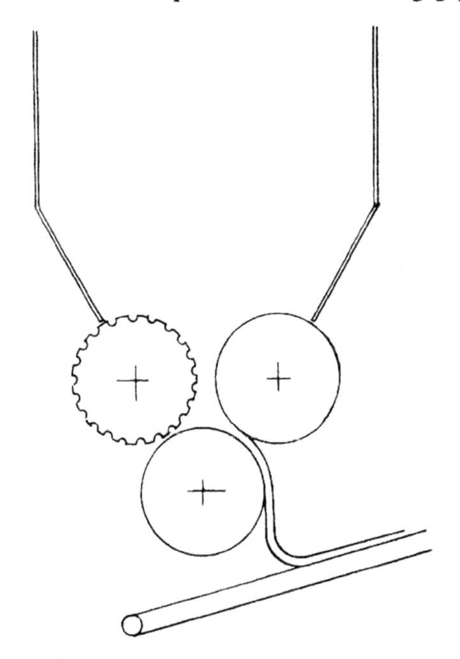

6 Sheeter with typical hopper shape.

extrusion of dough through the machine (that is, more dough passes through for a given number of revolutions of the rolls). Thus, for process control purposes, it is advisable to try to maintain a constant level of dough in the hopper and the best way to do this is via a pre-sheeter metering to the three roll sheeter. However, it is quite common practice to drop a whole mix of at least half a tonne of dough into the hopper at intervals. In this case a change in delivery rate from the sheeter must be expected as the hopper empties. In order to maintain precise control of the cutting machine it is necessary constantly to adjust the speed of the sheeter. This is not easy to achieve unless there is some form of automatic control sensor.

Where cutter scrap is fed back to the hopper of the sheeter other problems may arise. If there is a continuous feed of fresh dough to the sheeter it is quite satisfactory to run the scrap into the rear or front of the hopper for incorporation. However, where the dough is fed intermittently in large masses, provision must be made to allow a regular feed of scrap along with the fresh dough. This is usually done

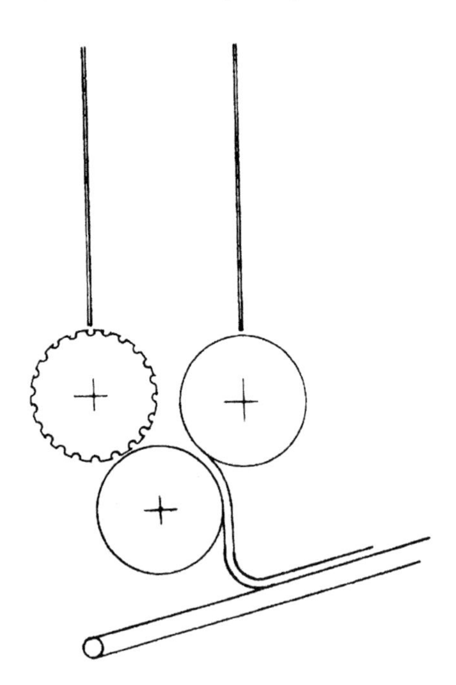

7 Sheeter with vertical sided hopper.

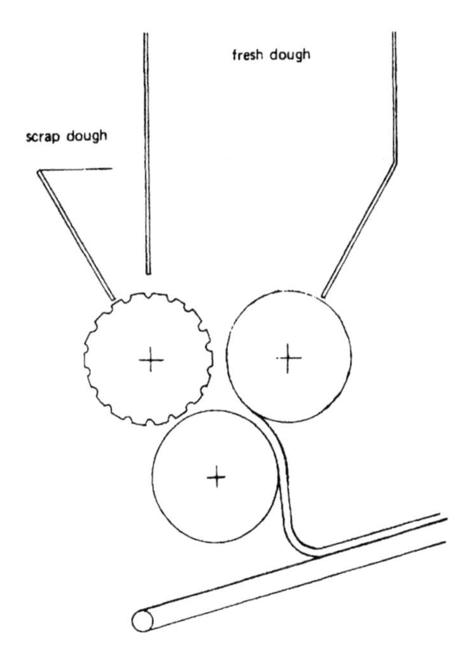

8 Three roll sheeter with pocket for cutter scrap dough.

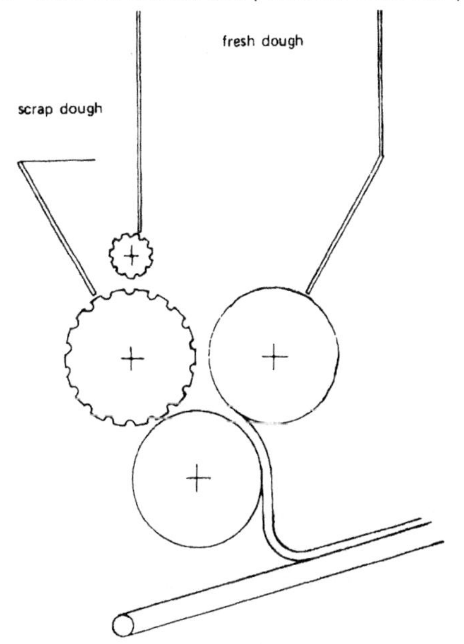

9 Three roll sheeter with roll to aid the metering of cutter scrap dough.

with the aid of a gap or small feed roll as shown in Fig. 8 and 9. A problem arises because the amount of scrap or the way in which it is taken away varies and some adjustment of the gap or height of the feed roll over the forcing roll is required – this is rarely provided and control is difficult or unsatisfactory.

It is usually possible to adjust the forcing gap and the gauging gap on a sheeter. Typically the former gap is about twice that of the latter. The greater the forcing gap relative to the gauging, the higher the compression on the dough and, obviously, the more work that is done on the dough. The efficiency of a sheeter for compaction of the dough is related to the ability to grip the dough and pull it into the compression chamber at the centre of the sheeter. The grooved rolls are designed to improve this gripping action.

The surface of the sheet as it emerges from the sheeter is of great importance to both the baked biscuit surface appearance and often to the degree of lift obtained in the oven. It is usual that a rough, rippled or holey sheet surface cannot be satisfactorily 'repaired' during subsequent gauging so the quality and integrity of the dough sheet from the sheeter is very important.

The dough sheet emerging from the sheeter is collected on a take-off conveyor which is driven by the sheeter and whose relative speed can be adjusted over a short range to ensure that the sheet lies well without being pulled from the sheeter. The conveyor takes the dough to the first gauge rolls.

5.2.3 Gauge rolls

Pairs of heavy steel rolls are used gradually to reduce the dough sheet thickness to that desired for cutting. Typically, there are two or three pairs although only one pair may be used for short doughs and more than three where very gentle reductions are necessary. As a rule of thumb the reduction in thickness at each gauge roll should be about 2:1, although ratios of up to 4:1 are commonly used.

Obviously the greater the ratio the more work and stress is put into the dough.

Usually the pair of gauge rolls are mounted vertically one above the other. Adjustment of the gap is made by moving one roll, sometimes the upper roll and sometimes the lower roll.

All gauge rolls should have instruments indicating the gap setting so that the machine can be changed or the settings recorded with accuracy. Often the gauges do not correspond well with the gap, but this is only a question of engineering maintenance for calibration.

Claims are made about the benefits of using rolls of various diameters ranging from about 150–300 mm and about different surface finishes. Both aspects are related to the tendency of the dough to stick to the roll as it emerges. There is also an effect on the smoothness of the surface of the dough. The roll surface can be rough, derived by sand blasting, polished (or even chromium plating) or special treatment with low friction coatings. The latter tend to have poor wear characteristics. Detailed investigations have shown that while these different surfaces have a marked effect on the dough release properties, they also affect the work done on the dough. This results in different 'spring' characters of the dough as it emerges from the gauge roll and also affects the surface appearance.

Dough emerging from a gauge roll is always of a slightly greater thickness than the gap it has come through. This is as a result of the elastic properties of the dough ('spring') and also because some extrusion, as compared with rolling, occurs through the nip.

Scrapers are provided to aid in the release of the dough from the rolls and also to keep the rolls clean. The aim is to cause the dough to adhere preferentially to the lower roll so that the dough follows the roll and then falls away or is eased off by a scraper before passing onto the take away conveyor. If the dough adheres to the top roll it is difficult to achieve a smooth transfer. It has been found that if there is a small differential in the speeds of the rolls the dough tends to adhere more to the faster roll. Speed differentials of up to 12.5% are tolerable.

Both sheeter and gauging rolls locate together between flanges fixed on the sides of one of the rolls. The flanges allow the dough to fill the roll right to the edges, thus ensuring a full width sheet without a ragged edge. Normally the flanges form part of the lower roll but as this makes it difficult to bring the take away conveyor close to that roll, further development has sited the flanges onto the upper roll instead. Release of the dough at the flanges can be a problem so that when upper flanged rolls are used there is a greater tendency for the dough to follow this roll rather than the lower roll. All of these problems are greatest when the dough sheet is thin, that is, they are most likely to be encountered at the final gauge roll pair.

Since tensions in the dough should be minimal and as constant as possible, it is necessary to maintain a full sheet by allowing a slight loop in the dough on the feed side of the gauge rolls and to have a similar loop at the discharge side. If the dough is pulled away from the rolls, tensions are produced that are worst at the edges. The optimum conditions are shown in Fig. 10.

Short doughs are non-elastic so careful attention must be given to the scraper position and the proximity of the take-off conveyor to prevent strong curvature at the discharge side of the gauge rolls. It is here that the top flanged roll configuration has maximum potential advantage.

Careful engineering adjustment is necessary for good performance of a series of gauge rolls. It is sometimes found that there is a tendency for overfeed at one side of the plant compared with the other. This may be due either to an uneven gap caused by non-parallel rolls or because the gauge roll pairs are not perfectly in line on the plant. To check for uneven gap across the roll, it is best to weigh precisely cut discs of dough taken from the sheet at each side. Testing the gap with feeler gauges when no dough is present will not show the effects of wear in bearings which will only be apparent when under load.

When machining harder or tight doughs it is frequently found that the dough is slightly thicker in the centre of the sheet than towards the edges. This is because some flexing of the rolls occurs under load. Before requesting engineering fixes for this problem see if it is possible to arrange that there is less reduction at the gauge roll by arranging that the sheet coming to it is thinner.

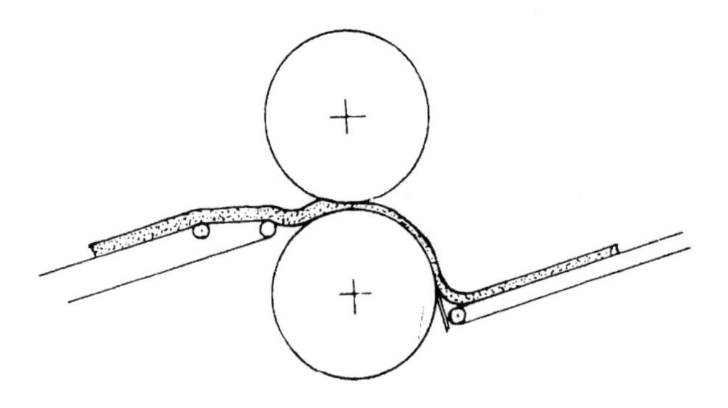

10 Optimum dough path through a gauge roll.

5.3 Dough piece cutting

5.3.1 Dough relaxation arrangements

The reasons for dough relaxation have been outlined above. Usually special provision is made for this only between the final gauge roller and the cutter. It may, however, be desirable to relax the dough more often in puff or other laminated types.

The dough is relaxed by allowing time and giving facility for unlimited shrinkage in the direction that the dough is travelling. In some older cutting machine plants the conveyor to the cutter was very long and the dough was slightly overfed onto it from the final gauge roll so that low ripples were formed. Before the sheet reached the cutter the shrinkage had absorbed the ripples so that a smooth sheet was available for cutting. The plant length required for this was rather long, especially on high speed lines, so an 'intermediate web' is now the more usual method to allow relaxation before the cutter. Dough is overfed onto this intermediate web to form ripples which may be quite large. From the intermediate web the dough is fed onto the cutting web. The speed of the intermediate web is adjusted so that the ripples in the dough are just taken out in time for a smooth sheet to be presented to the cutter.

This arrangement is reasonably satisfactory but this rippling frequently causes local transverse lines of stress, particularly in dry doughs, which are not completely removed before cutting. Also, it is easy to neglect the plant adjustment so that the dough is pulled off the intermediate web onto the cutting web thereby introducing some stress again. Deep rippling allows circulation of air under the dough which can reduce the adhesion of the dough to the cutting web at the point of cutting (see below).

The length of baked biscuits can be controlled by the amount of dough relaxation given before cutting. This means that it will not be desirable to give maximum relaxation before cutting if this part of the process is being used for length control. Clearly the amount of relaxation allowed depends on the amount of overfeed given to the intermediate web (or the cutting web if there is no intermediate web).

A typical arrangement of dough feed conveyors is shown in Fig. 11.

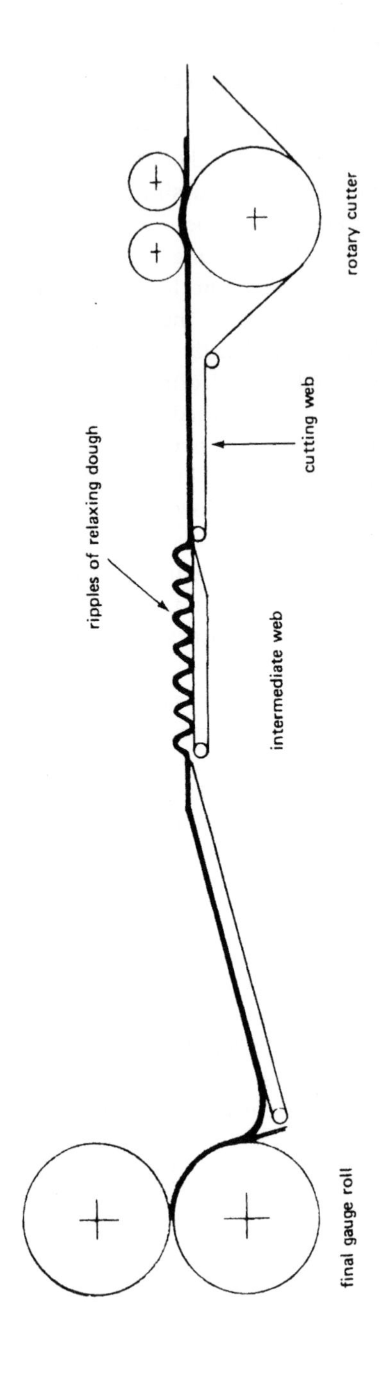

11 Typical arrangement of dough feed conveyors between the final gauge roll and the cutter.

5.3.2 Cutting

Older biscuit plants always employed reciprocating cutting machines. These use heavy block cutters which stamp out one or more rows of pieces at a time. The equipment needs to be strong and incorporates a swinging mechanism so that the dough sheet travels at constant speed and the cutter drops, moves with the dough, then rises and swings back before dropping again. Adjustment is provided to allow heavy or lighter cutting as the block falls. If the cut is too heavy there will be more damage to the cutting web, reducing its useful life.

In reciprocating cutting there are two basic procedures used depending upon whether merely cutting or embossing and cutting are required. Where simple cutting and dockering is needed, as for most crackers and hard sweet types, the cutting edges, docker pins and any type and decorative patterns are mounted on a base plate and a spring loaded ejector plate is located to move vertically around the fixed parts. When the cutter drops onto the dough the ejector plate is pushed back and as the cutter is lifted away the ejector pushes the dough to ensure it stays on the cutting web and does not stick to the cutter. If there is a tendency for the dough to stick to the ejector it is necessary either to dust the dough sheet lightly with flour, or to effect a little drying with a blast of air before the cutter.

When cutting and embossing is required, as for some short dough types, the ejector plate is replaced with an embossing plate. This plate is held back as the cutter drops and the plate is dropped to a predetermined position to press a deep pattern into the dough pieces. The cutting edges are then lifted away and the embossing plate is raised. This arrangement means that the dough pieces are determined first (which is important for controlling dough piece weight) then the surface relief is imprinted without loss of any dough by extrusion. It is unusual to have docker holes right through this type of dough piece, but if they are required, they are incorporated in the pattern cut in the embossing plate.

Reciprocating cutters thus require much heavy mechanism with many moving parts that need good lubrication and maintenance. They are often noisy, especially if the plant is run at high speed. Speeds of up to about 180 cuts per min can be achieved, usually less with embossing cutters than cutting only types. It is however unusual to run these cutters at more than 100 cuts per min. Two or more rows can be cut per drop of the cutter, but the more rows there are at once, the wider and heavier the cutter block becomes.

With the development of longer ovens (higher line speeds) and wider plants, it was necessary to consider improvements in the cutting arrangements. Rotary moulders have largely replaced embossing cutting and rotary cutters are now used widely for most other types.

Rotary cutters are of two types, those that employ two rolls, one immediately after the other, and those with only one roll. Dealing with the two roll type first, the principle is that the dough sheet, on a cutting web, is pinched between engraved rolls (mounted in series) and a rubber coated anvil roll(s) (see Fig. 12 and 13).

The first cutting roll dockers the dough, prints any surface pattern or type and thereby pins the dough onto the cutting web. The second roll is engraved with only the outline of the biscuit and cuts out the piece leaving a network of scrap (which has not been pinned down) similar to the reciprocating type of cutter. There is a facility for adjusting the rotational position of one roll relative to the other so that correct synchronisation between pattern and outline cut is possible. The pressure between each of the cutting rolls and the

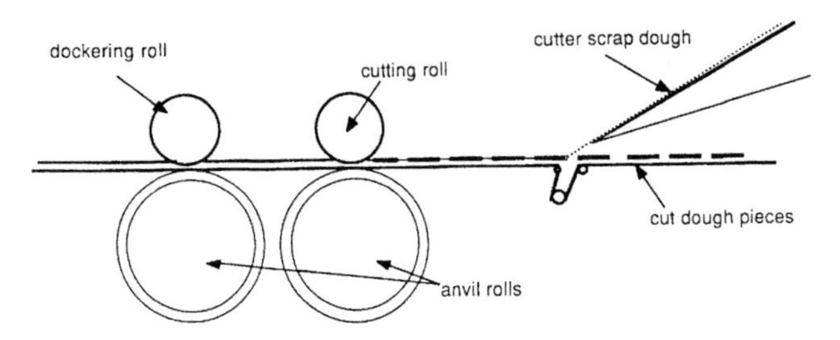

12 Typical rotary cutter arrangement (double anvil roll).

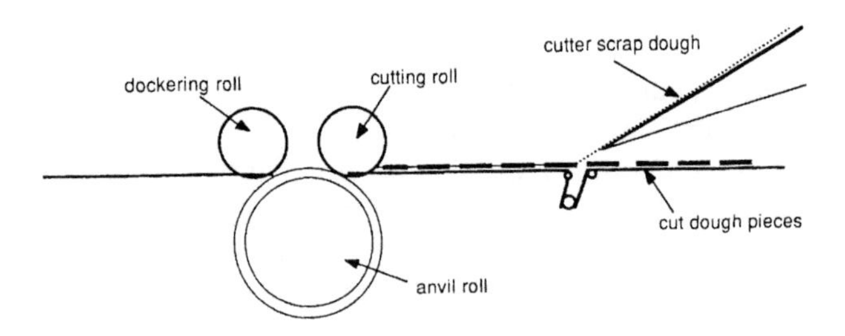

13 Typical rotary cutter arrangement (single anvil roll).

anvil can be adjusted independently but there is normally a quick release arrangement which lifts both rolls and maintains their relative positions so that production can easily be interrupted and resumed.

A fine adjustment is required for the height of the first roll to make sure that the dough is dockered and pinned down sufficiently, but the pressure must not be too great otherwise displacement of the dough both backwards to form a wedge and sideways affecting the thickness of the eventual dough piece may occur. The dough sheet thickness and hence the piece weight should be controlled at the final gauge roll not at the cutter. It is also important that the docker pins are not too long otherwise the cutting web may be damaged.

The cutting rolls are driven so that the speed relative to the cutting web can be adjusted slightly to affect the length of the cut piece and to assist release of the dough from the cutter. Setting the optimum relative speeds of the cutter rolls and the cutting web is critical. If the cutter speed is not right severe damage can be done to the cutting web.

A single roll rotary cutter achieves both dockering, pinning and outline cutting with only one roll. In most cases this works well and there is a saving in capital equipment, but there is a strong tendency to lift the dough piece from the cutting web because the pinning down facility is not independent of the cutting pressure. Cutting rollers are expensive so there should not be two where one would do. However, the frustration and disruption to production caused by malfunction indicates that the difficulties which may occur from using a single roll type are significantly minimised by using the two roll system.

There is no doubt that the performance of cutters, of either rotary or reciprocating types, is affected by the surface of the cutting web which carries the dough. The webs are usually woven cotton canvas or cotton/polyester mixture. Sometimes plastic coated webs are used. At all times a fine balance between good, but not excessive adhesion, of the dough to this web is sought. New canvas webs may have to be 'dressed' before use and webs which have dried out between production runs may also need some pretreatment. This treatment usually involves soaking the web in a liquid oil such as groundnut or soya oil, but sometimes a flour-water mixer is rubbed in to fill the pores and give a sticky surface. Some trial and error is involved. The dough itself dampens and conditions the web soon after production commences, but where a very dry dough is used or the bakery atmosphere is hot or dry, some additional dampening of the web may be needed throughout the production run. This can be achieved by a moistening (water) roller running on the return of the web under the cutter, or, alternatively, a steamer which lets steam condense on the web on its return run.

Plastic coated webs may offer too much adhesion due to their very smooth surface. These webs must be joined by splicing and heat bonding which is an ideal system when done well but one which requires special equipment. Canvas cutting webs are normally sewn by hand at the joint with polyester twine. The best webs are made endless but the design of the complete cutting section of the machine must be such that an endless web can be fitted easily.

Rotary cutting rolls are usually fabricated in bronze or gunmetal and modern automated engraving techniques allow high precision of shape and thickness of the patterns over the whole roll surface. Moulded plastic 'cups' which are attached to a roll surface are also a standard now. The moulded cups are produced from a single die and are all identical. The cost of the fabrication of a cutting roll with cups is not very different compared with the cost of an engraved roll but as it is possible to have an unlimited number of spare cups from the same mould, savings in the event of wear or damage are very favourable.

5.3.3 Cutter scrap dough handling

Ideally the network of dough not included in the cut pieces will adhere less firmly to the cutting web than the pieces. It can therefore be lifted and gently pulled upwards onto a scrap return conveyor (see Fig. 13). If the dough is very weak it may be necessary to use supports to aid the lift onto this scrap return web. These supports are called 'fingers'. These can only be used where the dough pieces are in straight lines with clear lines of scrap where the fingers can be located. Scrap fingers cannot be used where the pieces are cut in a staggered arrangement as is usual for round or oval biscuits.

A web 'break' arrangement may be provided just at the scrap lift position (see details in Fig. 14 and 15). This involves a downward loop in the cutting web which allows the dough to be peeled off at a sharp nose piece and thus be released at the crucial moment. Pieces which are peeled and then replaced on the cutting web in this way tend to transfer better to the panner or oven band. A web break should only be used when absolutely necessary as the extra nose piece and strong flexing of the web reduces the life of the web and its joint and also makes web tracking more difficult.

The scrap dough, having been lifted, is carried away on a full width conveyor whose speed can be controlled. Much of the success of good scrap dough removal relies on controlling the speed of the take away web.

The scrap dough may be returned as a full width network to the front of the sheeter hopper, or it may be collected and sent back on a narrow conveyor to be distributed across the back of the sheeter hopper or, less commonly, back to a bin or to a mixer. Whichever is the arrangement, it is important to ensure that this scrap dough is spread evenly in the sheeter to optimise distribution in the new dough. It is worth watching that edges of the new sheet are not rich in scrap as this will have an adverse effect on the quality of biscuits produced.

A particular problem with scrap reincorporation with fresh dough can occur when the full width of the dough sheet is lifted from the cutting web at start up or when cutting problems are encountered. If the amount of scrap dough involved is high it is better to collect the scrap in a bin and not to return it to the sheeter. Use of scrap dough collected in this manner is a matter of company policy.

If the quantity of scrap is normally high a two roll sheeter may be used to make a more or less continuous layer of scrap dough which is fed beneath the fresh dough from the main sheeter. The advantage of this arrangement is that the scrap can be metered and its placement is assured.

Scrap dough is always of different consistency from the fresh dough. This may be because it is cooler, having been exposed as it has passed along the cutting machine. In some cases the scrap dough is taken through a heater in a humid atmosphere as it is returned to the sheeter. Theoretically this is desirable but care must be taken to

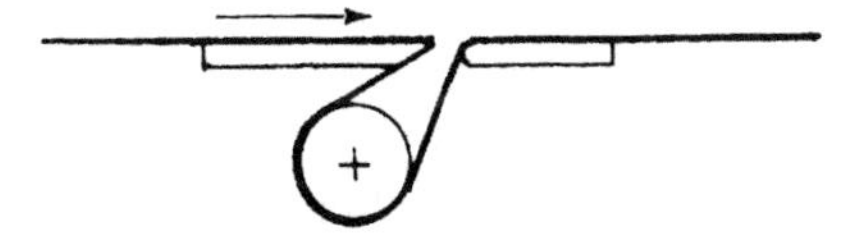

14 Web break with two nose pieces.

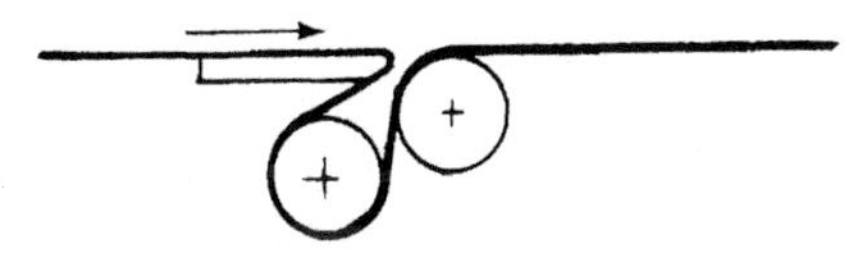

15 Web break with one nose piece and a roller.

ensure that mould growth is not allowed to develop in this warm area. Frequent inspection and cleaning are required.

5.4 Dough piece garnishing

Having removed the scrap dough, a surface dusting of sugar, salt or nuts, etc. may be applied to the pieces before they are placed onto the oven band. Systems for this operation always include a means of recovering the excess material which falls between the pieces or rolls off. The recovered material can be reused.

At this point, as an alternative to dusting, a milk or egg wash may be applied. This is normally done with a revolving brush although there are some cases where a spray is used. Care must be taken that as little as possible of the liquid spills onto the web as this will make the surface very sticky and necessitate continuous washing and drying. Egg or milk that gets onto the underside of the dough pieces will bake very dark and will look like dirt on the bottom of the biscuits.

5.5 Panning onto the oven band

It is not usual to run the cutting web right through to the oven band. Normally there is a transfer onto a panning web and it is on this web, or another intermediate conveyor, that the garnishing or surface washing takes place.

The finished dough pieces are transferred (panned) onto the oven band. The panner unit has several functions. It allows acceleration of the pieces off the cutting web which may be useful if some spatial separation, etc. is required and it is usually capable of swivel so that it can follow the line of the oven band as it tracks slightly from side to side. In this way the dough pieces are always deposited away from the extreme edges of the oven band.

The transfer of pieces from the panner to the oven band must be adjusted carefully so that the spacing is optimum and they lie down smoothly. Badly panned pieces may be distorted in shape after baking. Spacing of the dough pieces on the oven band is a critical process control point. If the pieces are spaced too widely the oven is not being used to optimum capacity. Ideally the spacing should be as close as possible but the spacing in the line of travel should be equal to the lateral spacing that was determined by the cutter.

5.6 Start up and shut down of the cutting machine

5.6.1 Start up

Before dough is placed in the sheeter hopper or dough feed machine and at least 10 min before production is due to commence.

Start these checks at the oven end of the cutting machine:

- Check that the oven band is running and the production speed is correct.
- Check that any garnishing machinery is charged with ingredient and ready to run.
- Check that the correct cutter is fitted.
- Check that the cutting head is disconnected (reciprocating cutting) or that the cutting rolls are not in contact with the cutting web.
- Check that the cutters are clean.
- Check that there are no objects lying on the any of the conveyors of the machine or in the dough hopper.
- Check that all the catch trays are in place.
- Set the sheeter and gauge roll gaps as specified on the process audit chart for the biscuit to be made.
- Start the cutting machine.
- Set the speeds of each machine as specified on the process audit chart and check that each machine is running.
- Check that all the fabric conveyors are tensioned and are tracking correctly and that the seams are in good condition.
- Check that dough will be ready for use at the appointed time.
- Note the time that dough feeding must commence to allow dough pieces to enter the oven according to the production schedule. (In some cases a short trial run of dough may be required to check the phasing of the cutter rolls or the skip of the reciprocating cutter and to test the oven heat.)

As production commences

- Check that the oven is ready for production, i.e that it is at the correct speed and temperature, and that the baker is in attendance.
- Give instructions for the dough to be tipped into the sheeter or dough feed hopper.
- Follow the dough as it runs down the plant and adjust the speeds

of the feeds into the gauge rolls, as necessary, assist the doughs away from each gauge roll onto the take off conveyors.

- Start air blasts and flour dusters as necessary.
- Adjust the dough relaxation before the cutter if necessary.
- Run the full dough sheet under the cutter and onto the scrap return web.
- Engage the cutter and adjust the speed, pressure and registration of rolls as necessary.
- Adjust the scrap lift so that dough pieces pass forward to the oven.
- Start the garnishing machine, as necessary.
- Check that the spacing of the dough pieces onto the oven band are correct by adjusting the speed of the panner.
- Signal to the baker that product is entering the oven.
- Check the dough piece weight against the process audit 'standard'.
- Write down the time that production started.
- Recheck that the plant is running correctly, paying particular attention to the feed into the first gauge roll.
- Check that there is not too much pressure at the embossing, first cutter roll, which could be affecting the weight of the dough pieces.
- Recheck that the dough piece weight is as expected and be ready to receive information from the baker that the biscuits are of the correct weight and shape. Adjust as necessary and signal to the baker when the adjustment has been completed.
- Check that the garnishing machine is running correctly and check garnished dough piece weights if necessary.
- Check that the cutter scrap is being incorporated correctly in the sheeter.
- As dough which includes scrap passes down the plant, adjust the feeds to the gauge rolls as necessary and recheck the dough piece weights.

5.6.2 Shut down

At the end of the designated production time

Company policy will dictate whether the plant should run to an exact time or whether production should end when dough runs out of the sheeter.

- As the last of the fresh dough runs low in the sheeter hopper divert the cutter scrap into a tub.

- As soon as the last of the dough runs under the cutter, completely release the cutter away from the cutting web and stop the cutter when the last pieces have been run onto the oven band.
- Stop the garnishing machine.
- Signal to the baker that the last of the dough pieces has entered the oven.
- Write down the time that production finished.
- Stop the cutting machine and commence cleaning operations.

5.6.3 Stops during production

Stops due to faults or difficulties on the cutting machine

- Release the pressure of the cutter on the cutting web before stopping the machine.
- Signal to the baker that there is an interruption of production.
- Try to rectify the fault, call for assistance quickly if this seems necessary.
- If the stop has been short, restart and continue as normally.
- If the stop has been prolonged and the condition of the dough has deteriorated, upon restarting, before engaging the cutter, send all the old or dry dough into a tub rather than back to the sheeter. Only when the plant is running well should the cutter be re-engaged and the cutter scrap returned to the sheeter.
- Inform the baker that production has recommenced.
- Write down the time of restart, the duration of the stop and the reason for it.

5.7 Troubleshooting

5.7.1 Dough handling and sheeting

1 Dough sheet is not complete from the sheeter.
  - insufficient dough in the hopper
  - dough bridging in hopper, try feeding the dough so that the level in the hopper is kept lower. Is the dough consistency too hard and dry? Is it possible to use a softer dough?
  - the take away from the sheeter is too fast and is breaking the sheet
  - the dough is too rich in scrap, the hopper level was too low or the mixing with fresh dough is not good enough

auging.
ugh before gauging look further

robable that there is an overfeed
the dough into the nip of the rolls
he width of the rolls.
olls is full, an underfeed situation
e dough and cause an incomplete
ls
l before gauging look further back

erging sheet.
er inclusion is caught behind the
briefly to allow it through
on the surface of the upper roll.
per roll and increase the pressure

aper on the upper roll.
fat. The problem is not a serious

nents

he dough as it is transferred onto

transfer is pulled a little
dough down evenly onto the web

varying suggesting irregular

gh are constant before and after
rlier in the plant can affect the
roll
ssing pressure is not too great
l of the cutting rolls
intermediate web are variable in
direction of the plant.
ebs at the point where dough is
te web, it should be the same on

5.7.4 Cutting

5.7.4.1 Reciprocating cutting

Before making any adjustments check the dough piece weight is correct. This, in effect, will mean checking the baked biscuit weight.

1 Dough pieces are sticking to the ejector plate/embossing plate.
- check that the embossing pressure is not too great
- dry the dough surface with air or flour
- increase the adhesion of the dough to the cutting web by wetting the web or placing a roller onto the web immediately after its transfer onto the cutting web (see Section 5.7.3 point 2)
- check that the cutter is clean and the edges of the cutting rim are not damaged

2 There is difficulty in getting enough cut pressure to release the dough pieces from the surrounding scrap network.
- placing a piece of web under the cutting web at the point where the cutting takes place may give a little more resilience to the cutting web allowing better cutting.

5.7.4.2 Rotary cutting

Before making any adjustments check the dough piece weight is correct. This, in effect, will mean checking the baked biscuit weight.

1 Changes in the setting of the final gauge roll do not affect the dough piece weight.
- this will be because there is too much pressure at the first roll, the embossing roll. Every time an adjustment is made to the dough sheet thickness, the setting of the cutter should be checked to ensure that there is no gauging action at the first roll

2 Difficulty is experienced in getting correct or consistent registration of print and cut between embossing and cutting rolls
- the dough is probably sticking to the first roll, rising up and between the two rolls.
- reduce the pressure of the first roll and if this does not give enough imprinting attend to the adhesion of the dough to the cutting web (see Section 5.3.2)
- skim the dough surface with an air blast or flour duster

- As soon as the last of the dough runs under the cutter, completely release the cutter away from the cutting web and stop the cutter when the last pieces have been run onto the oven band.
- Stop the garnishing machine.
- Signal to the baker that the last of the dough pieces has entered the oven.
- Write down the time that production finished.
- Stop the cutting machine and commence cleaning operations.

5.6.3 Stops during production

Stops due to faults or difficulties on the cutting machine

- Release the pressure of the cutter on the cutting web before stopping the machine.
- Signal to the baker that there is an interruption of production.
- Try to rectify the fault, call for assistance quickly if this seems necessary.
- If the stop has been short, restart and continue as normally.
- If the stop has been prolonged and the condition of the dough has deteriorated, upon restarting, before engaging the cutter, send all the old or dry dough into a tub rather than back to the sheeter. Only when the plant is running well should the cutter be re-engaged and the cutter scrap returned to the sheeter.
- Inform the baker that production has recommenced.
- Write down the time of restart, the duration of the stop and the reason for it.

5.7 Troubleshooting

5.7.1 Dough handling and sheeting

1 Dough sheet is not complete from the sheeter.
   - insufficient dough in the hopper
   - dough bridging in hopper, try feeding the dough so that the level in the hopper is kept lower. Is the dough consistency too hard and dry? Is it possible to use a softer dough?
   - the take away from the sheeter is too fast and is breaking the sheet
   - the dough is too rich in scrap, the hopper level was too low or the mixing with fresh dough is not good enough

2 Dough sheet has a rough surface.
   - insufficient compression in the sheeter
   - too much compression resulting in too much work and stress in the dough. Try increasing the forcing gap setting. Is it possible to use a softer dough? Is the scrap responsible for the problem and if so is it being incorporated evenly?

3 The feed from the sheeter is variable at a constant speed.
   - the level of dough in the hopper is varying too much or the consistency of the dough is varying, possibly due to temperature or dough age
   - the scrap dough incorporation is variable
   - the scrap dough has dried out too much and is causing reincorporation problems.

5.7.2 Gauging of dough

1 The feed of dough at each side is not similar.
   - check the thickness of the dough sheet after the preceding machine. The gauging rolls may not be parallel to each other on this or the preceding machine
   - check tracking of web carrying dough to the gauge roll, is it taking the dough to one side?
   - if this is a persistent problem, check alignment of the sets of machines

2 Dough sticks to the upper roll.
   - is the reduction excessive and could it be reduced?
   - is the dough overfeeding into the nip?
   - is the scraper set correctly to clean the roll?
   - is the dough homogeneous? Are there patches which seem richer in syrup or fat that are causing the sticking?
   - if the dough seems variable report to the mixing department
   - use an air blast or flour duster on the dough surface immediately before the gauge roll to skin the surface and reduce the stickiness
   - usc a light roll to hold the dough down over the scraper and onto the take off conveyor

3 Dough sticks to the lower roll and is difficult to take off at the scraper.
   - check that the scraper is set correctly to keep the roll clean
   - pull the dough off onto the conveyor
   - use dough that is not so soft

4 Dough surface is rough after gauging.
   - if the dough surface was rough before gauging look further back up the plant
   - attend to the feed as it is probable that there is an overfeed situation causing rippling of the dough into the nip of the rolls
5 Dough sheet is not full across the width of the rolls.
   - check that the feed into the rolls is full, an underfeed situation will cause the rolls to pull the dough and cause an incomplete sheet to emerge from the rolls
   - if the dough sheet was not full before gauging look further back up the plant
6 A torn line appears in the emerging sheet.
   - a hard piece of dough or other inclusion is caught behind the nip of the roll, open the gap briefly to allow it through
7 A film of material builds up on the surface of the upper roll.
   - examine the setting of the scraper roll and increase the pressure if possible
8 Oily fat is dripping from the scraper on the upper roll.
   - place a drip tray to collect the fat. The problem is not a serious one.

5.7.3 Dough relaxation arrangements

1 Air is becoming trapped under the dough as it is transferred onto the cutting web.
   - arrange that the dough at the transfer is pulled a little
   - place a light roller to hold the dough down evenly onto the web at the transfer
2 The length of the biscuits is varying suggesting irregular relaxation before cutting.
   - check that tensions in the dough are constant before and after each gauge roll. Variations earlier in the plant can affect the relaxation after the final gauge roll
   - check that the cutting/embossing pressure is not too great causing variations in the speed of the cutting rolls
3 The ripples in the dough on the intermediate web are variable in angle or are not transverse to the direction of the plant.
   - check the height of the two webs at the point where dough is transferred onto the intermediate web, it should be the same on both sides of the plant.

5.7.4 Cutting

5.7.4.1 Reciprocating cutting

Before making any adjustments check the dough piece weight is correct. This, in effect, will mean checking the baked biscuit weight.

1 Dough pieces are sticking to the ejector plate/embossing plate.
- check that the embossing pressure is not too great
- dry the dough surface with air or flour
- increase the adhesion of the dough to the cutting web by wetting the web or placing a roller onto the web immediately after its transfer onto the cutting web (see Section 5.7.3 point 2)
- check that the cutter is clean and the edges of the cutting rim are not damaged

2 There is difficulty in getting enough cut pressure to release the dough pieces from the surrounding scrap network.
- placing a piece of web under the cutting web at the point where the cutting takes place may give a little more resilience to the cutting web allowing better cutting.

5.7.4.2 Rotary cutting

Before making any adjustments check the dough piece weight is correct. This, in effect, will mean checking the baked biscuit weight.

1 Changes in the setting of the final gauge roll do not affect the dough piece weight.
- this will be because there is too much pressure at the first roll, the embossing roll. Every time an adjustment is made to the dough sheet thickness, the setting of the cutter should be checked to ensure that there is no gauging action at the first roll

2 Difficulty is experienced in getting correct or consistent registration of print and cut between embossing and cutting rolls
- the dough is probably sticking to the first roll, rising up and between the two rolls.
- reduce the pressure of the first roll and if this does not give enough imprinting attend to the adhesion of the dough to the cutting web (see Section 5.3.2)
- skim the dough surface with an air blast or flour duster

  - moisten the surface of the cutting web, there may be a unit for doing this as it is commonly needed at the start of production.
3 The scrap dough network is sticking to the cutting roll and being pulled up by it. This can be a particular problem with short doughs where the thickness of the sheet is relatively thick.
  - Reduce the pressure of the first roll to a minimum, it could be that dough is being displaced and is therefore thicker in the scrap dough area
  - Consider fitting nylon lines looped around the second roll and also around a bar fitted in front of the cutter. The lines will increase the release of the scrap dough (see Fig. 16)
4 The dockering and print on the dough piece surface is not clear and there is evidence of a tearing around the docker holes.
  - the relative speed of the embossing roll is incorrect, probably too fast
5 Dough pieces are sticking in the cutting roll.
  - try adjusting the relative speed of the cutter to the cutting web
  - attend to the surface condition of the cutting web to make it slightly more adhesive
6 Dough pieces are too long or too short.
  - adjust the relative speed of the cutter to the cutting web.

5.7.5 Dough piece handling after cutting

1 The dough pieces do not transfer well from either the cutting or panning web or do not cross over a web break point satisfactorily.
  - can the nose piece of the web be brought nearer to the receiving conveyor?

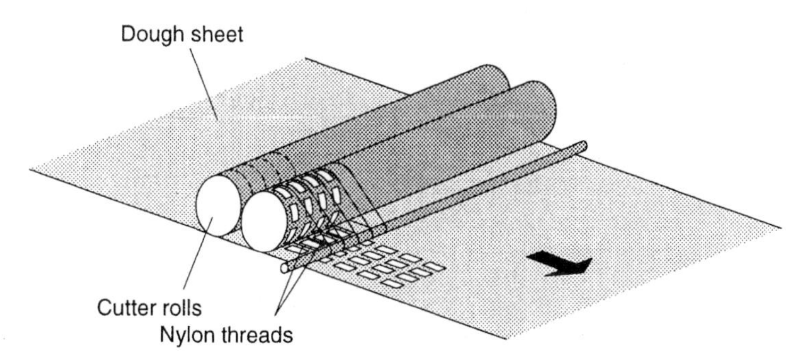

16 Cutting roll with nylon threads to aid release of scrap dough.

  - adherence is too great on the cutting web and the transfer does not allow a smooth peeling away from the web. Refer to the points made on cutting in Section 5.3.2 and troubleshooting for cutting, Section 5.7.4, and consider whether less adherence of the dough to the web, compatible with satisfactory cutting, can be adjusted
2. The scrap web does not separate well from the cut pieces.
   - the cut is not complete, increase the pressure of the cut
   - the cutting web is too adhesive, probably too wet
   - use the web break system, if available, to release the dough from the cutting web before the scrap take off
3. The dough pieces transfer better on one side of the plant than the other.
   - check the relative levels of the conveyors at each side, the gap between them should be absolutely parallel, decide whether a greater or smaller gap is better but remember that space must be allowed for the web seam to pass through the gap
   - the cutting web is too adhesive possibly due to uneven wetting if this is being used.

6 Laminating

6.1 Principles and techniques of laminating

The purpose of laminating is to build up a layered structure by rolling the dough then folding it and turning it, through 90° at least once before gauging to the final thickness for cutting. This develops the gluten and gives a delicate flaky structure in the baked biscuit. In many cases another material, such as fat, is introduced between the layers to encourage further separation of the layers during baking.

Originally laminating was done manually using a reversing brake but now automatic laminators are used.

Dough lamination is now used principally for cream cracker and puff doughs but any developed (hard) dough can be laminated with benefit. If semi-sweet or savoury cracker doughs are made using sodium metabisulphite or enzyme to modify the gluten it is not necessary to laminate as straight sheeting and gauging will produce satisfactory biscuits.

Pizza dough is basically a bread dough. Because of the gas retention properties required, the gluten must be strong and relatively non-extensible. Sheeting of this dough is difficult but laminating improves it considerably.

Cream cracker dough is somewhat like pizza dough in that it has a bread-type recipe and is fermented with yeast. It is difficult to sheet smoothly but it is usually laminated not only to create a clearer dough but also to allow the introduction of flour and fat mixture ('cracker filling dust') designed to keep the dough layers separated a little during baking. The fact that some cream cracker manufacturers use very little 'cracker dust' tends to show that it is the dough sheet/ stress relieving function of the laminator that is of most importance to the biscuit structure.

By introducing a lot of fat between the dough layers a flaky structure is produced and this is the characteristic feature of puff biscuits. The fat may be introduced as knobs into the dough prior to sheeting, or as a continuous layer after a good dough sheet has been formed. In either case, gauging and laminating is needed subsequently to build up the number of dough layers which appear as very thin flakes in the baked biscuit or pastry.

A critical feature of **puff dough** production is the plasticity of the fat or fat/flour mixture that is used. It is necessary for the consistency of the fat and dough to be similar so that the fat does not break through the dough layer or, alternatively, squeeze out. In order that the fat used does not have a large high melting point tail above body temperature but is sufficiently firm and plastic in consistency it is normal to arrange that the dough is cold, usually not more that 18°C.

'Laminated' means that the dough has been reduced to a thin sheet and this sheet has then been piled up into a number of layers, with or without the introduction of another material between the layers, and the pile has been further gauged to a final dough sheet for cutting. The term 'gauging' is used to mean the reduction of sheet thickness by rolling between a pair of rollers.

6.2 Types of laminator

Having outlined the reasons for laminating, let us now consider the various types of automatic laminator. The following types can be identified:

1 Vertical with continuous lapper and one sheeter (see Fig. 17). This usually is composed of a three roll sheeter with cutter scrap incorporation, two or three gauge rolls, cracker dust spreader on part of the sheet and a zig-zag lapper (see Fig. 18) capable of building up about 10 or 12 layers. The advantage of this type of laminator is the continuous smooth action of most parts, but the disadvantages are:

   (a) There are stresses introduced in the laminated dough at the edges due to folds.
   (b) The exposure of top and bottom of the sheet on successive Vs of the folded dough (this can mean that the scrap dough is alternately exposed if incorporated in one side of the sheeter) (see Fig. 18).
   (c) The cracker duster is intermittent in action and must be synchronised with the lapper. The filling dust is normally between only every other lamination (see Fig. 19).

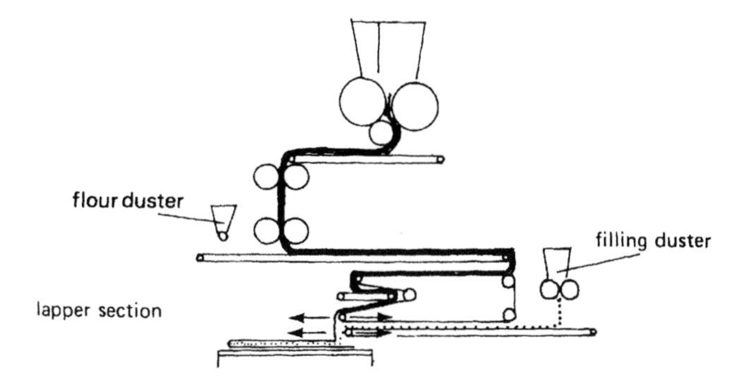

17 Vertical laminator with continuous lapper.

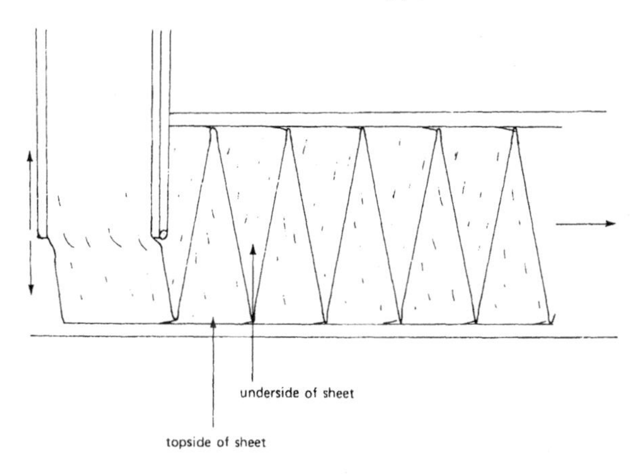

18 Formations of laminations on a continuous lapper.

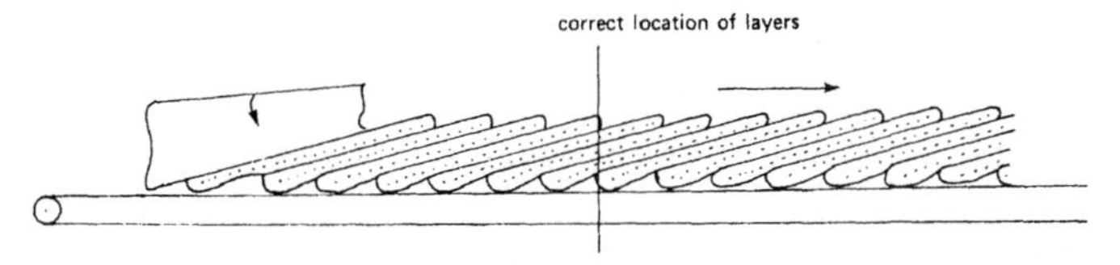

19 Formation of four double laminations with a continuous lapper. The filling dust is shown between each double layer.

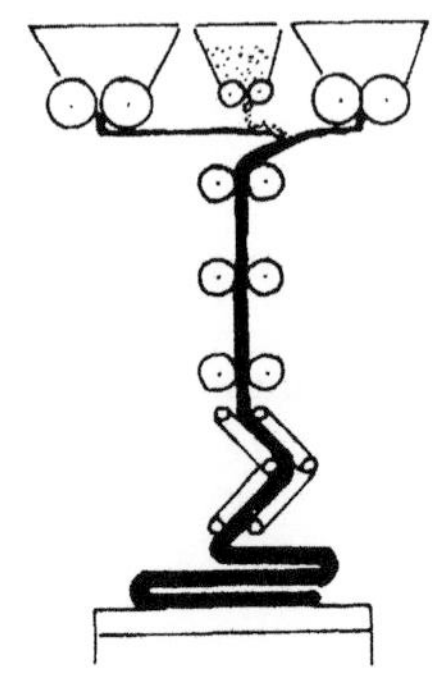

20 Vertical laminator with two sheeters and a continuous lapper.

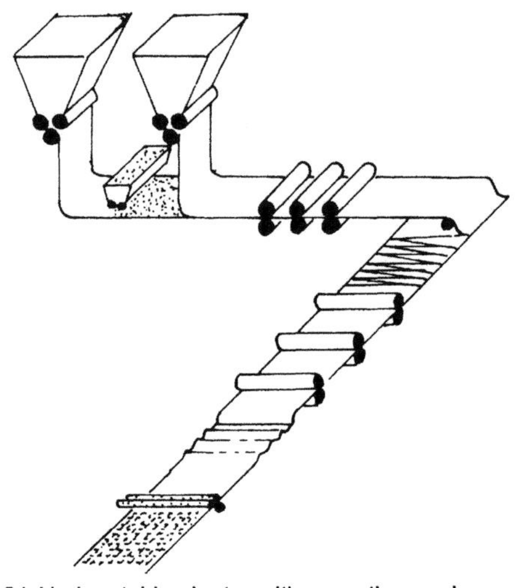

21 Horizontal laminator with a continuous lapper.

2 Vertical with continuous lapper but with two sheeters (see Fig. 20). Here two-roll sheeters make sheets and filling dust is incorporated prior to subsequent gauging. The advantage of this type would appear to be that the filling can be spread continuously over the full width of the sheet, but disadvantages are the same as for the previous type of laminator combined with the fact that by using two roll sheeters, poor sheets may be formed and it is these sheets that have to 'hold' the filling.
3 Horizontal laminators (see Fig. 21). These are similar in performance to the vertical types, but the sheeting and gauging

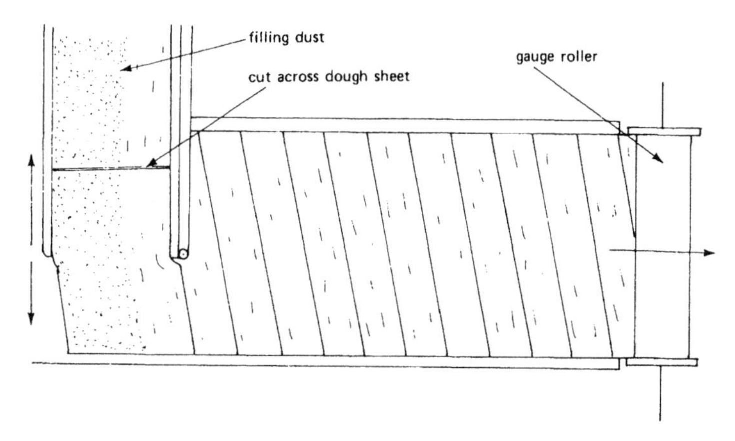

22 Formations of laminations with a cut sheet laminator. Dough is deposited only on the backward stroke of the lapper.

(also the cracker dust filling) occurs on units spread out horizontally before the lapper, more like conventional biscuit cutting machines. The disadvantage of this type is that the whole machine takes a lot of floor space because a right angle bend in the line of plant is required where the lapper is sited. The advantage is that more than one lapper can be used if required, introducing a second 'turn' to the dough. It is usual to use two three-roll sheeters in horizontal laminator systems and cracker dust or more fatty fillings can be added between the two sheets (that is, on the lower sheet of dough).

4 Cut sheet laminators. Both the vertical and horizontal laminators described above are of the continuous sheet folded type, but either can be supplied with a cut sheet lapper. This means that a dough sheet is laid down only as the lapper retracts; only the top of the dough sheet appears on the pile of laminated dough (see Fig. 22). The advantages are that little or no stress is involved at the edges since no folding occurs and that an in-line laminating arrangement is an option. The disadvantages are that a more complicated laminating action is needed to cut and lay down the sheets and cut edges may expose the filling onto the subsequent surface of the dough. This is particularly important where a heavy fat filling is involved.

The vertical laminators were introduced to save floor space and by reducing or eliminating canvas webs between gauge rolls much machinery was saved. However, machining dough involves stresses in the gluten network and these stresses can only be relieved with time. The close proximity of the components in the vertical laminators allows much less time for natural stress relief than the horizontal configurations. Various types of vertical laminators have been designed with resting webs between gauging stations. These machines begin to approach the complexity and cost of the horizontal types.

In all laminators the width of the laminated pile of dough can be controlled by attention to the length of the stroke of the laminating section. The number of layers formed can be altered by adjusting the relative speed of the conveyor that takes away the laminated dough. The faster this conveyor runs, the fewer the number of layers.

6.3 Process control during laminating

Many laminators have a bulk rather than a metered feed of dough into the sheeter hopper. This means that the discharge of sheeted dough is rather irregular due to the height of dough in the hopper. This problem was discussed for sheeting in Section 5.3. Thus there is the need for continual surveillance of speed here to maintain an even feed into the first gauge roll.

As the lapper builds up the layers of dough it is important to make speed adjustments so that exactly the same number of laps is present at all times. If a variable thickness occurs, as shown in Fig. 23, the feed to the next gauge roll will be uneven and irregular stresses will be set up in the dough on gauging to a final thickness. The best check on whether the laps are placed correctly is to observe the dough as it emerges from the next gauge roll. If it shows bands of crushed or stretched dough associated with the edges of each lap, appropriate opening or closing of the space between each lap can be made by changing the speed of the laminator relative to the laminated dough web.

It is important for the quality and shape control of the baked biscuit that the amount of filling introduced between the laminations is uniform. Malfunctioning of the filling duster will result in uneven weights and lift of biscuits and may be enough to affect the coloration and eating qualities. There should be a standard ratio of filling dust to dough and a procedure should be devised to enable checks to be made that this standard is maintained. Unfortunately, it is not always simple to collect the delivery from the duster to permit checks on weights to be made.

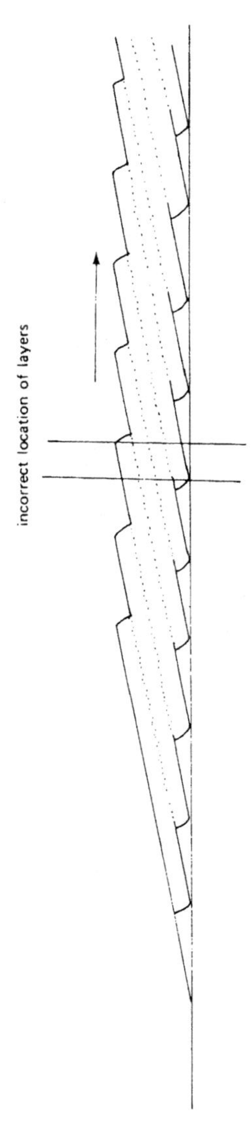

23 Formation of incomplete four laminations with cut sheets showing position of filling dust between the layers.

The number of laminations required to give an optimum quality of biscuit should be decided by experiment. Too few laminations will give a ragged flaky structured biscuit that may split open along the lines of the laminations and too many will result in excessive crushing during gauging prior to cutting leading to a biscuit that has poor thickness development. The latter effect can be reduced if a thinner dough sheet is laminated.

Poor quality edge-lane biscuits are often observed in laminated products. Careful observations at the gauging after the laminator will usually indicate the reasons. If, for example, the width of the laminated dough is too great, crushing will occur at the edges; if too narrow, some pulling will be apparent. It is simple to make adjustment to the length of the stroke of the laminator carriage.

6.4 Start up and shut down of the laminator

6.4.1 Start up

Before starting production with the laminator:

- Check that there are no objects lying on the any of the conveyors of the machine.
- Set the sheeter and gauge roll gaps as specified on the process audit chart for the biscuit to be made.
- Fill the cracker dust distributor if this is to be used.
- Check that the cutting machine is running at the correct production speed.
- Start the machine.
- Set the speeds of each section of the laminator as specified on the process audit chart and check that each machine is running.
- Check that all the fabric conveyors are tensioned and are tracking correctly and that the seams are in good condition.
- Set the speeds of each part as specified on the process audit chart and check that each part is running correctly.

When dough feeding starts:

- Check that the cutting machine and oven and are ready for production.
- Start the cutting machine as the speed of the laminator is referenced from this.
- Follow the dough as it runs down the laminator and adjust the speeds of the feeds into the gauge rolls, as necessary.

- Adjust the speed of the layering carriage to give the exactly correct placement of the folded dough, or cut sheets, on the take-off conveyor.
- Start the cracker dust machine, if appropriate, and check that it is delivering the correct amount of filling dust.
- After the cutter scrap has reached the sheeter make periodic checks on the feeds to the gauge rolls and the placement of the laps.

6.4.2 Shut down

At the end of production:

- Stop the laminator when all the dough has run through.
- Clean the sheeter rolls.
- Empty the cracker dust distributor.

A laminated biscuits mechanism chart is shown in Fig. 24. See Manual 4, *Baking and cooling of biscuits*, for more information on baking.

6.5 Troubleshooting problems with laminating

1 The baked biscuits lack thickness.
  - Check biscuit weight.
  - Try decreasing the gauge roll gap of the next to last gauge roll (the penultimate roll), not the final gauge roll the one before it. This will result in less crushing work being done at the final gauge roll which may increase the lift in the oven.
  - Could be adjusted by changing the baking profile, see Manual 4 on *Baking and cooling of biscuits*.
  - Could be a dough crushing effect after laminating and before cutting.
    - check that the placement of the laps is accurate otherwise crushing could be occurring at the first gauge associated with uneven thickness of dough
    - look for overfeeding at each of the gauge rolls after the laminator
  - Could be that there are too many laminations for the strength of the gluten in the dough.
    - try one less lap of dough at the lamination point. This will involve first slowing the laminator relative to the take-off

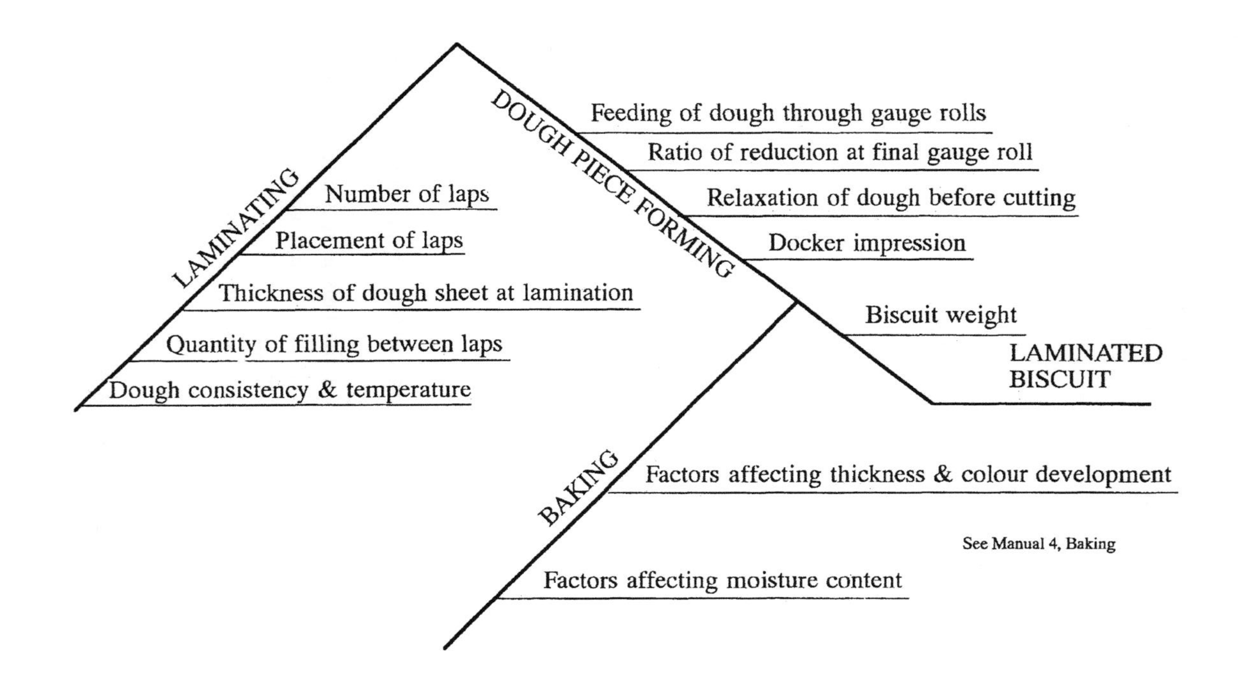

24 Laminated biscuit mechanisms.

conveyor, then speeding the take-off conveyor when the new set of laminations arrives at the gauge roll. Allow the plant to settle and see the effect on baking

2 The baked biscuits are too thick.
   - Check biscuit weight.
   - Try increasing the gauge roll gap of the next to last gauge roll (the one before the final gauge roll). This will result in more work being done on the gluten at the final gauge roll which will reduce the lift in the oven.
   - Could be adjusted by changing the baking profile, see Manual 4 on *Baking and cooling biscuits*.

3 The biscuits tend to delaminate, split open.
   - Could be due to too much filling dust or flouring of the dough before lamination.
     - check the amount of filling dust being used
     - increase the pressure on the dough at the cutter to pin the layers of dough together more firmly
     - increase the number of laminations.

4 The biscuits show variable baked colour and thickness.
   - Try to establish a pattern of the variation.
     - is it related to the exposure of top and bottom sides of the dough sheet as a result of laminating?
       If so the effect may be due to the incorporation of cutter scrap on one side of the dough sheet. The cutter scrap will be richer in fat from the filling dust than the fresh dough and will colour more during baking. The solution is to reduce the amount of filling dust or the amount of fat in the filling dust.
     - is it related to biscuit weight variations?
       If so try to decide why the biscuits could be varying in weight. Is the feed of dough at the gauge rolls uneven, sometimes pulling giving thinner sheets?
     - is it related to the way in which cutter scrap is being incorporated at the sheet? Perhaps there are periods when the sheet is much richer in cutter scrap than at others. If so try to regulate the incorporation of the cutter scrap better.

7 Rotary moulding

7.1 Principles of rotary moulding to form dough pieces

A rotary moulder is a machine commonly in use for producing biscuit dough pieces from short doughs. The dough is forced into moulds which are the negative shape of the dough pieces complete with patterns, name, type and docker holes. The excess dough is scraped off with a knife bearing upon the mould and thereafter the piece is extracted onto a web of cotton canvas or other fabric.

Although short doughs may be sheeted, gauged and cut with an embossing type of cutter the advantages of moulding are:

- it is not necessary to form and support a dough sheet.
- difficulties and control of gauging are eliminated.
- there is no cutter scrap dough which must be recycled.

The last is of great significance as short dough toughens as it is worked and gauged. In a rotary moulder all the dough has a similar history, there is no cutter scrap that has to be reincorporated. The shape of the mould allows a much more intricate pattern outline than cut dough and can give hollow centres to the pieces if required. Large or very small dough pieces can be moulded but there can be extraction difficulties if very thick dough pieces are moulded.

7.2 Rotary moulding machine

7.2.1 General description

Figure 25 shows a typical rotary moulding machine. Roll A is known as the forcing roll. It is usually made of steel and has deep castellations, in various patterns, designed to hold a blanket of dough. The roll is driven so that dough from the hopper (H) is drawn down into the nip against roll B. Roll A may or may not be adjustable in a horizontal direction. Roll B is the moulding roller. Typically it has a similar diameter to roll A but it has a smooth surface into which

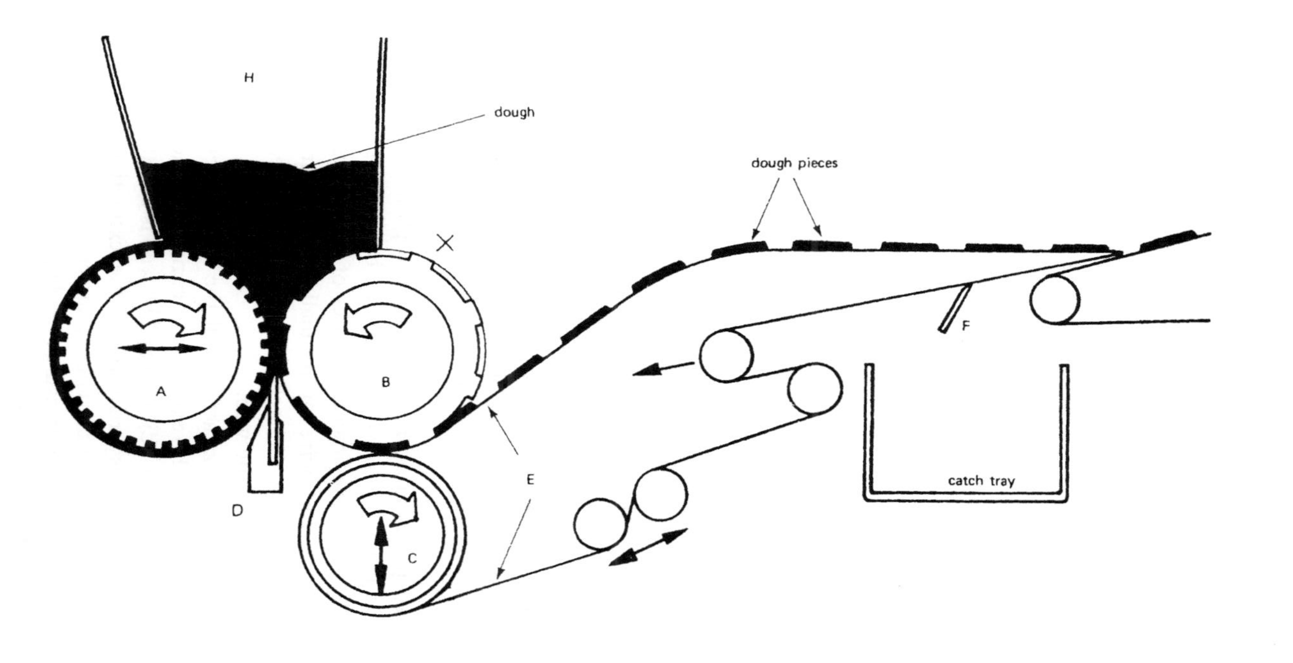

25 Cross section to show the parts and action of a rotary moulder.

are engraved, or inset, moulds to form the shape of the dough pieces. Typically the roll is made of bronze or gunmetal (a malleable alloy of copper, tin, zinc and sometimes lead) which is suitable for delicate engraving. However, if plastic insert moulds are made the roll may be of steel. In both cases the roll, which is the same width as the biscuit plant, is usually a tube mounted on an axle. It is important that this tube and axle is rigid and does not flex under the pressure of dough created in the nip with the forcing roll. The moulding roller is driven as shown so that dough is forced into the moulds in the nip. Its position is fixed.

Bearing on the moulding roll is a blade of steel known as the scraper (D). The tip of this blade is below the centre line of rolls A and B where maximum dough pressure is exerted in the nip. Dough which has been forced into the mould is sliced off and the excess runs down the scraper and is pressed into the blanket of dough which adheres to the forcing roll. The scraper knife may be adjusted in its position on the moulding roller but the ways and means of achieving this vary in different moulders. Ideally it should move tangentially to the surface of the moulding roll.

Roll C is the extraction roll. It has a thick rubber coating over a steel centre and around it passes the extraction web (E). By adjusting the position of this roll in a vertical direction the extraction web can be pressed against the moulding roller. It is driven in the direction shown and the dough pieces are pulled out from the moulds onto the extraction web. The hardness of the rubber on the extraction web is fairly critical and with time and use this changes. The rubber coating will have to be replaced at intervals to maintain optimum efficiency of the rotary moulder.

The dough pieces are carried to the nose piece where they are peeled off and panned either on to the oven band or an intermediate web. In order to aid the smooth transfer of dough pieces from the extraction web a thin wire or a small diameter metal shaft is sometimes used near to the web nose piece to prevent the dough piece going round with the web.

On its return to the extraction roll the web passes over a web cleaning scraper (F) which removes any remaining traces of dough. The tension of this web is adjustable. The web is seamless requiring a design that enables easy removal of the extraction roller. The extraction web will need to be replaced at regular intervals. The life of this web depends on the pressures needed to run a dough and on whether the moulds have docker pins.

To form different biscuits the scraper must be moved away from the moulding roll which is then exchanged for a different one. This change is straightforward and quick, but as the rolls are heavy, lifting tackle is always required. Great care must be taken not to knock or drop the moulding roll while it is being moved as it is expensive and the metal which is relatively soft is easily damaged.

The position of the scraper tip is typically between 3–11 mm below the axis of rolls A and B. The scraper knife warrants some further consideration. In order to slice the dough the blade should be as sharp as possible, but as there is a great pressure of dough in the nip, the tip of the blade should not be so thin that deflection towards the moulding roller can occur or else it will cut into the metal. The scraper is sprung so that the tip always runs against the roller otherwise dough may pass behind it tending to force it away and causing it to engage with the forcing roll and cause much damage. As some flexing of the scraper knife tip is inevitable, the docker pins in the mould must be fractionally lower than the level of the mould edge or they will be damaged by the blade.

When changing the height of the scraper blade it is important to move it as nearly as possible in a tangential direction relative to the moulding roll surface.

7.2.2 Formation of a dough piece

Dough is placed in the hopper and the machine is started. Dough trapped at the nip is churned and worked to force it through the nip.

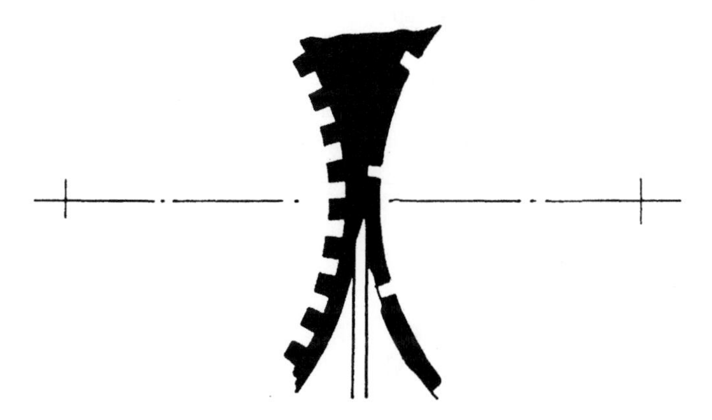

26 Scraper knife in high position to show how dough becomes extruded past it.

This churning may toughen the dough, but the toughening effect is less if the dough from the mixer has been allowed to stand for at least 30 min before use. The level of dough in the hopper should be maintained at a minimum so as to reduce pressure differences at the nip and excessive working of dough and also to minimise the chance of bridging of the dough. The dough is typically of a slightly firmer consistency than would be the case if the dough was to be sheeted.

The dough is pressed against the forcing roll and into the moulds on the moulder. The scraper knife slices off the dough level with the top of the mould and presses the excess against the forcing roll to form a blanket which revolves with the forcing roll. Depending upon the position of the knife some dough may be forced round behind the knife effectively to overfill the mound (see Fig. 26 and 27).

The dough piece passes to the point where extraction is achieved. This is the place where greatest difficulties can occur. For example, the dough piece may stick preferentially to the mould and therefore not be extracted or there may be a squeezing action that causes the piece to be wedged in cross-section and some dough extruded behind the piece in a 'tail' on the extraction web (see Fig. 28).

The surface of the extraction web must be sufficiently rough or sticky to effect good adherence of the dough piece, but the adhesion must not be so good that subsequent peeling at the transfer from the web is difficult. The surface and type of web is important and many have been tried. A thin web is usually not rough enough and a thick web will not go round a sharp nose piece. A sharp nose piece is

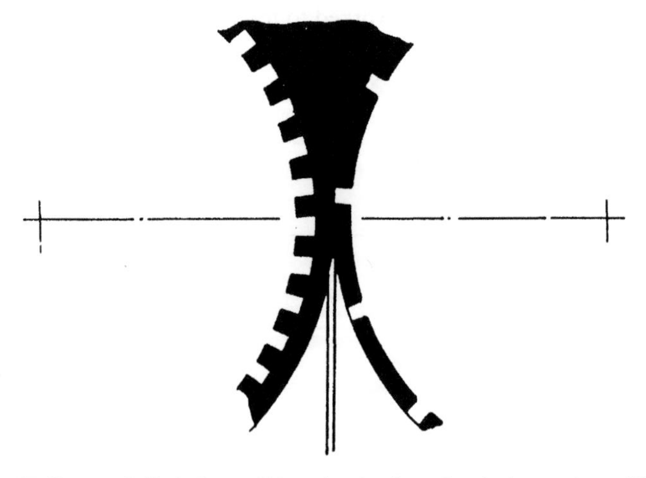

27 Scraper knife in low position showing how dough does not overfill the mould.

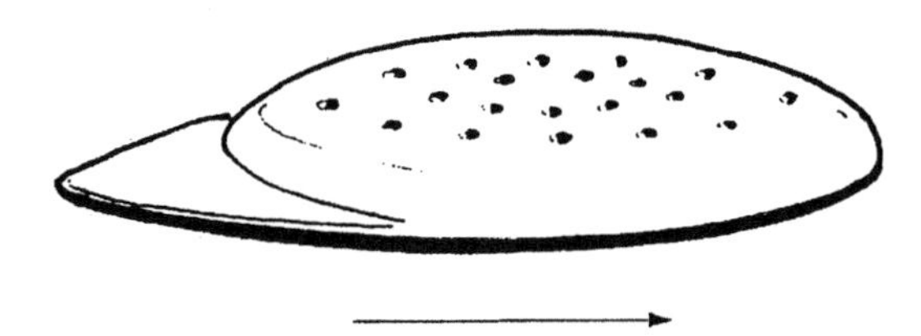

28 Dough piece with a tail of dough.

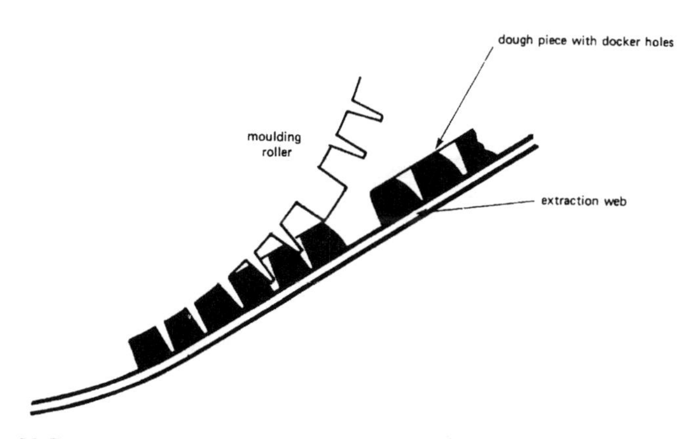

29 Extraction from the moulds to show problems associated with thick dough pieces.

needed to cause a peeling away of the dough pieces. The internal surface of the moulds is important in both shape and smoothness with regard to ease of removal of the dough piece. If the mould edges are too steep or the pattern too deep or intricate, extraction will be difficult. If the mould is very deep with docker pins the physics of the change in direction from rotation to linear at the extraction point can give difficulties for removal of the dough piece. The pins will have to tear through the dough or will hold it in the mould (see Fig. 29).

Low friction coatings such as PTFE (polytetrafluoroethylene) are useful for lining the moulds but they soon wear off. Plastic insert moulds are effective for aiding dough piece release. In these plastic moulds docker pins, where necessary, are usually made from bronze for added strength. If the release is made too easy from the moulds the pieces may fall out in the period between the scraper and the extraction point. This problem is accentuated by the fact that there is

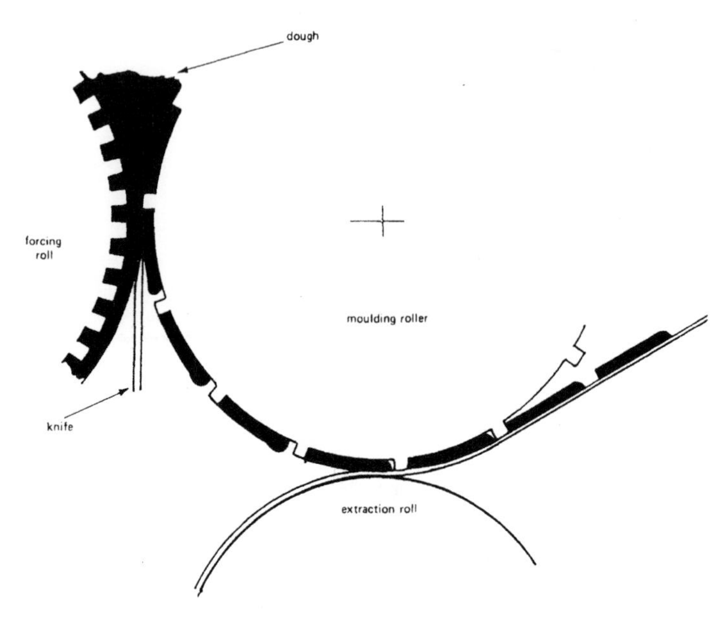

30 Detail of rotary moulder action to show effect of dough drag on the scraper

a drag from the knife that tends to pull the front of the dough piece away from the mould causing it to curl (see Fig. 30). This also affects the shape as the piece is pressed onto the extraction web and, incidentally, is the main reason why extensible or toughened doughs cannot be successfully rotary moulded.

Other aids to extraction include heat applied at point X (see Fig. 25) or a light spray or brushing with release oil at the same place.

Extraction is effected by the web being pressed against the dough held in the mould. The slightly soft surface of the extraction roller allows the web to be pressed into the mould. This has two effects. First, the dockers pass right through the dough to the web and, second, a slight excess of dough is extruded outside the limits of the mould. The rolling action means that the dough is extruded predominantly at the rear of the piece and this forms a 'tail'. Some tail is almost inevitable, however, the tail will be excessive if:

- There is high extrusion round the scraper blade, associated with high pressure in the nip or a high blade position;
- There is too much pressure at the extraction point due to the extraction roll being set too high; or

- The rubber surface of the extraction roll is too soft allowing too much dough to escape at the back of the mould.

The pressure between the moulding roll and the extraction web should be the minimum for satisfactory dough piece extraction.

7.3 Dough feeding to a rotary moulder

Feeding dough to a rotary moulder is more critical than to a three-roll sheeter. There are two important considerations. First, the level of the dough in the hopper of the moulder should be kept at a minimum to avoid bridging and this involves supplying the dough in small pieces. Second, because the moulding roll is expensive and made of a relatively soft metal, great care must be taken that no pieces of metal pass into the hopper with the dough.

Thus dough fed to a rotary moulder typically comes through a pre-sheeter, which is of two-roll design. It passes to the hopper of the moulder on a full width conveyor and then passes through a kibbler before it falls into the hopper. The kibbler is a set of rotating fingers that break the dough into pieces of no more than $50\,mm^3$. The conveyor from the pre-sheeter passes through a metal detector and there is a rejection arrangement so that any dough found to contain a metal fragment is taken away before it can reach the moulder and damage the moulding roll or the knife.

There is typically a dough level sensor in the hopper of the moulder and the pre-sheeter and conveyor are started and stopped to meter the dough to the moulder.

7.4 Dough piece weight control

Compared with a cutting machine there is less opportunity for weight control using a rotary moulder. However, there are a few techniques, which are interactive and can be used to adjust the dough piece weight a little. First, if the forcing gap is adjustable the smaller the gap, the higher the pressure and the higher will be the dough piece weight due to slight change in dough density. If there is a separate speed control for the forcing roll, increasing the speed will also result in an increase in pressure in the forcing gap and the dough piece weight will increase a little. As described above, the position of the scraper knife affects the amount of extrusion of dough around it. Thus the higher the knife position the greater the amount of dough in the mould. Conversely the lower the knife position the less dough there will be in the mould. At the extraction point, the higher the pressure the lower will be the weight and the greater the tail. Related to this, the softer the extraction roll surface the lower the dough piece weight and the greater the tail (and wedging!). The consistency of the dough may have some effect on the dough piece weight. Often drier doughs give higher dough piece weights. Figure 31 gives a diagrammatic representation of the effects of dough piece weight.

The methods used to **increase** the dough piece weight from a rotary moulder may be summarised:

- Increase the height of the scrap position.
- Decrease the forcing gap.
- Reduce the extraction pressure.
- Reduce the water content of the dough.

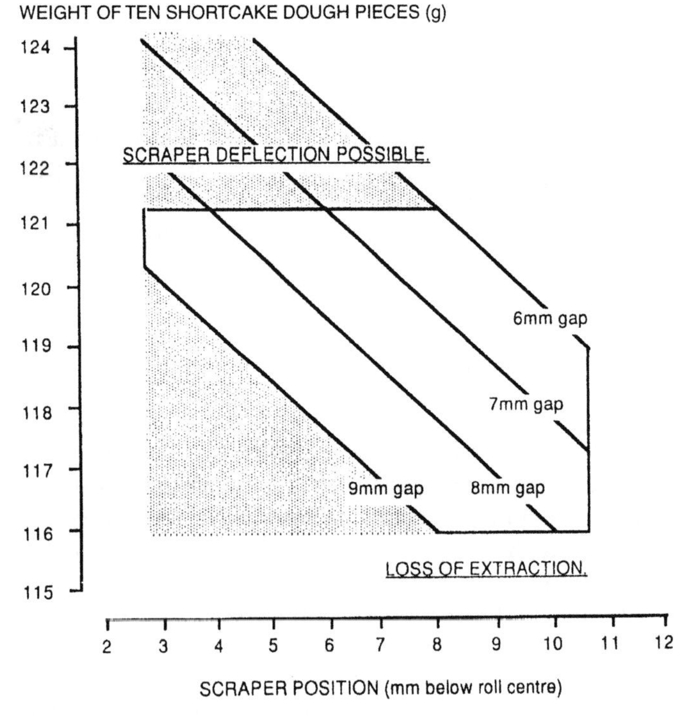

31 Diagrammatic example of effects on dough piece weight of changes on rotary moulder settings

The opposite actions will result in some reduction of dough piece weight.

It is recommended that trials are made with each moulding roll to establish how much adjustment to dough piece weights can be achieved by making the changes listed above.

With time the surface of the moulding roll slowly wears so the depth of the moulding cups becomes a little less. There will therefore be a general reduction in dough piece weight with time.

7.5 Start up and shutting down of a rotary moulder

7.5.1 Starting up

Before placing dough in a moulder:

- Check that the moulding roll is very clean and that there is no dough in the castellations of the forcing roll.
- Start the machine, set speed, the forcing gap and scraper position according to the process audit chart for the biscuit to be made.
- Check that the extraction pressure is off, or is just sufficient to cause the extraction web to run if the extraction roll is not separately driven.
- Ensure that the extraction web is in a good condition to extract the dough pieces, dampen slightly with water if necessary. Ensure that the dampness is even over the whole of the web.
- Check that the web is tensioned ready for running.
- Check that the metal detector in the dough feed is switched on.

After starting the feed of dough to the moulder:

- Allow the blanket of dough to form on the forcing roll.
- Gradually increase the extraction pressure until dough pieces are formed across the whole web. Aim for a minimum pressure and repeatedly check for this.
- Check that the transfer of the dough pieces from the extraction web onto the next conveyor is smooth and satisfactory.
- Check the dough piece weight.

7.5.2 Shutting down

At the end of production:

- Run the dough out of the hopper.
- Take off the extraction pressure.

- Clean out the dough from the forcing roll, a special tool may be needed for this, and clean the roll as well as possible.
- Set the scrapper away from the moulding roll in the 'parked' position.
- Lift out the moulding roll and take it away for thorough cleaning, preferably with hot water and leave it to cool and dry for later use. NOTE that the roll should be cleaned while the dough in the moulds is soft. It is much more difficult to clean it after the dough has dried out. It is not satisfactory to dry clean a roll without removing it from the machine as films of sugar syrup will not be removed and these will hinder extraction at the next production run.
- Clean dough from the web scrapper tray and from under the machine.
- Check the condition of the scraper blade. (N.B. Be VERY careful when cleaning or inspecting this knife as it can be very sharp.) If it is found to be even slightly damaged immediately report for engineering attention.

7.6 Troubleshooting

7.6.1 Difficulties when extracting dough pieces from the moulding roll

- Exceptionally high extraction pressure is required to obtain satisfactory extraction and this causes long tails.
  - check that the dough has stood for at least 30 minutes after mixing. This reduces the stickiness significantly
  - if there is provision for adjusting the differential speed of the extraction roll and web compared with the moulding roll try to change the speed **slightly**
  - try **slightly** moistening the extraction web
  - try increasing the dough piece weight by using the techniques described above in Section 7.4
  - try using a slightly softer dough
  - try mixing the dough a little longer. A slightly toughened dough will extract better
  - stop the machine and remove the roll for thorough cleaning (do not try to refit it while it is still hot)
  - try applying heat at point X on the moulding roll (see Fig. 25). A warm air blower may be satisfactory.

- Extraction is better on one side of the machine than the other.
  - adjust the differential roll pressure as the extraction roll may not be parallel with the moulding roll. Engineering assistance may be required for this.
- Extraction from selected moulds is poor.
  - clean the moulds with compressed air (wear safety goggles to protect eyes)
  - check the moulds which are holding the dough for signs of damage at the edges
  - check that the extraction web is not damaged
- Extraction is intermittently poor.
  - check the feed of dough in the hopper, is it running too low or is it bridging?
  - check condition of extraction web and extraction roller. The latter may have deposits of dough on it that should be cleaned away
  - check the age of the dough. Fresh dough will be more difficult to extract than dough which has stood for at least 30 min
- Extraction at the front edge of the dough piece is often poor.
  - increase the height of the scraper position to increase the pressure that the dough is forced into the moulds. This makes the whole dough piece more cohesive and easier to extract

7.6.2 Difficulties with dough piece weight and shape

- The dough pieces are too low in weight.
  - try one or more of the techniques listed in Section 7.4
- The dough pieces are strongly wedged (thicker at the back than the front or the other way round).
  - try reducing the dough piece weight, see Section 7.4
  - try using a drier dough
- The dough pieces seem to be varying in weight across the machine.
  - stop the machine, slacken off the extraction web and clean the surface of the extraction roller and check that the rubber is not damaged
  - check that the extraction web is not damaged. Change to a new one if it is damaged

- The biscuits exhibit transverse cracks or are fragile in a transverse line.
  - this may be caused by the dough pieces cracking before they reach the oven. The extraction web should follow a gentle curve away from the extraction point to the nose piece. If it is taken over a sharp point the dough pieces, particularly if they are thick or formed in very dry dough, will tend to crack. These cracks will impair the strength of the biscuits after baking (see Fig. 32).
  - the dough pieces may be cracking as they are transferred from the extraction web (this may be a particular problem if the dough pieces are very thin). Adjust the relative speeds of the take-off web and rotary moulder
- The dough pieces are incomplete, being short of dough at the front of the piece and often with excessive tails.
  - the dough is too tough and is pulling back at the scraper knife. Reduce the mixing time or stand the dough for longer before placing in a pre-sheeter
  - certain doughs that are too low in fat may not be suitable for rotary moulding as there is too much gluten formed during mixing
  - try raising the scraper position
  - the dough is too sticky. Try standing the dough for longer before use or reducing the dough water level.

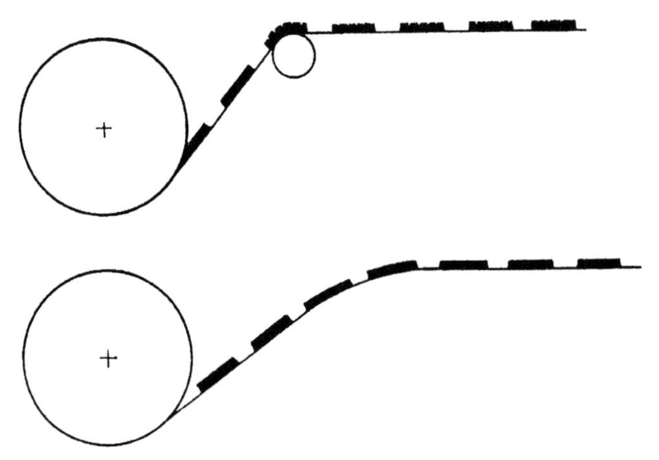

32 Above: Inflection of dough pieces causes permanent cracks. Below: Optimum arrangement for extraction web to minimise damage to dough pieces.

7.6.3 Problems with the dough blanket on the forcing roll

- The dough will not form a blanket on the forcing roll and it falls off:
  - the dough may be too dry
  - increase the pressure of the dough in the forcing gap by reducing the gap size by raising the knife position or by increasing the speed of the forcing roll if this roll is separately driven.

8 Extruding, wire cutting and depositing

8.1 Principles and techniques of extrusion and depositing

Extrusion and depositing constitute some of the simplest means of making dough pieces. The technique involves forcing dough through holes in a die plate. Where the dough is very fluid and probably has been aerated in the course of mixing, it is known as a batter. Thus the dough is extruded rather than sheeted or moulded. Depositing is a form of extrusion. The mechanism of these two means of dough piece forming are not distinct from one another. However, the machinery that handles soft short doughs is normally different from that which deposits a dough that is so soft it is pourable. The firmest doughs are often wire cut and these may have a similar consistency to soft rotary moulder doughs. Wire cutting makes it possible to form pieces from more sticky doughs and dough containing coarse particles, such as nuts or oatflakes, which cannot be successfully rotary moulded. In extrusion and depositing dough which has been pressurised either by means of rollers (short and soft doughs) or a pump (sponge batters) is forced through orifices. The two types will be dealt with separately.

8.2 Dough extrusion machines

Most machines consist of a hopper over a system of two rolls which force the dough into a pressure/balancing chamber underneath. The rolls may run continuously or intermittently and may be capable of a short period of reverse motion to relieve the pressure and cause a suck back at the dies or nozzles at the base of the pressure chamber. Thus dough can be forced continuously or intermittently out from the pressure chamber.

The machine spans the width of the plant and is usually situated over the oven band. In the case of certain drier, wire cut doughs and rout types (see Section 8.3) which are subsequently cut into lengths before baking, the machine is over a normal canvas conveyor and not the oven band. Dough pieces formed on a conveyor may be spaced out as they are transferred onto the oven band.

The dies of the extruding machine are usually about 70 mm above the oven band, or take away conveyor, but it is possible to adjust this as necessary. The size of the extrusion is determined by the size of the dies and the rate of extrusion is adjusted by the speed of the forcing rolls. The rate of the extrusion is affected by the consistency of the dough and the pressure in the chamber behind the die plate. The rate may also be affected by the head of dough in the hopper. Some deferential pressure in the chamber behind the dies across the machine may cause uneven extrusion but ways of compensating for this will be discussed below.

8.3 Formation of dough pieces by wire cutting, rout press and co-extrusion

Figure 33 shows the arrangement of a wire cut machine. For wire cutting, the dies are about 70 mm above the band or conveyor, but provision is usually available to change this gap if necessary by raising or lowering the band. Dough is extruded through a row of dies (of any desired size or shape) and a frame bearing a taut wire or

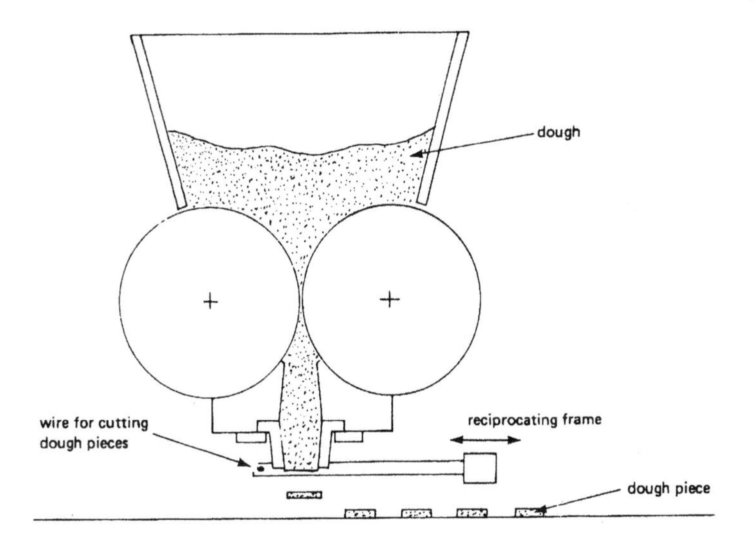

33 General arrangement of wire cut machine.

blade, strikes across the base of the die holes 'cutting' off the extruded dough at intervals. The dough pieces then fall onto the band or conveyor. The wire may cut in a forward (direction of the oven band) motion or more usually in the opposite direction. In any case, the cutting stroke is close to the die and the return stroke is lower, away from the dies so that it does not touch the dough which is continuously being extruded. The pieces may fall off straight or they may turn over before they reach the band. It does not matter whether they turn or not, but all must do the same all the time! It is here that difficulties are often encountered because sticky, coarse textured dough may not always fall as expected. Some control of the fall of the dough piece is possible by adjusting the height of the dies above the band and also by attention to the position of the wire as it passes through on the cutting stroke. Every effort must be made to achieve a uniform position on the band so that the pieces will experience similar baking conditions.

Sometimes a blade with sharp edge or even a fine serrated edge is used in place of the wire. Sometimes the wire or blade may be vibrated, in a horizontal plane, to improve the 'cutting' action. The best conditions for a particular dough are, unfortunately, found principally by trial and error. Having found the good settings, record them in terms of machine calibration.

The speed of wire cut machines is not high, rarely exceeding 100 strokes per min, although the number of rows can be increased either by having a double row die, if the pieces are fairly small, or by having two machines synchronised to deposit alternate rows of pieces.

Dough extruding from a round die may be distorted somewhat as the wire or blade 'cuts' across it. This is particularly common if the dough contains coarse pieces or dried fruit that drag on the wire. The shape of the baked pieces will be affected, but may be not as much as expected.

Wire cut dough pieces always have a rough surface which is not completely lost during baking. A 'home make' look is accentuated by the fact that the outline after baking is often irregular. The nature of extrusion means that the dough pieces are thicker in the centre than at the edges. A constriction in the die can reduce this effect.

It is, of course, possible to have different die shapes across the band and even, by dividing the hopper, to use two or more different doughs at once. This allows variety production for packs of assorted biscuits, but the dough recipes and piece weights should be calculated carefully to achieve optimum baking of all types.

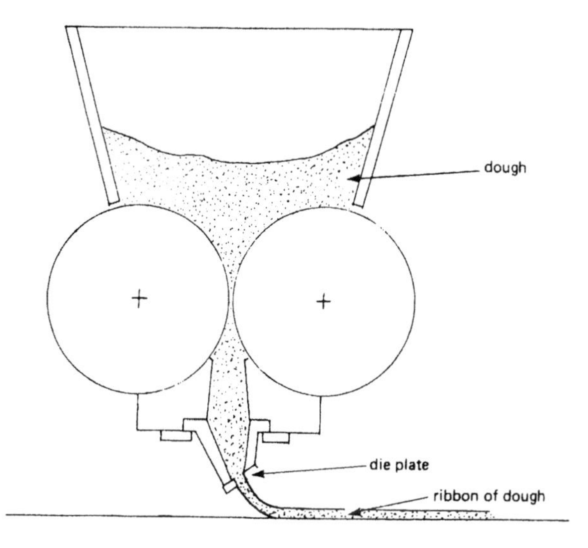

34 General arrangement of rout press machine.

When the machine is used in a continuous mode, without wire cutting pieces, the products are called bar or rout cookies. The die plate is usually inclined in the direction of the extrusion so that transfer of the dough ribbons onto the conveyor is as smooth as possible (see Fig. 34).

For rout cookies the dies are typically of asymmetric shape to give a smooth base and ribbed upper surface. The ribbons of extruded dough are usually guillotined into short lengths before baking, but alternatively, some form of cutter may be used on the oven band after baking. Depending on the character of the dough in terms of coarse ingredients and consistency, the bar cookies may have smooth or rough edges and surfaces.

It is possible to elaborate on the bar cookie by arranging twin pressure chambers under separate hoppers so that one dough is extruded within the other. This coextruding arrangement is used in the production of, for example, fig bars. The fig (or other centre fill material) paste is fed from one hopper and dough is fed from the other (see Fig. 35). The filled tube so produced may be cut before or after baking as described above.

The disadvantage of this simple type of co-extruder is that it is difficult or impossible to have a cutting device that seals the filling within the dough when the pieces are cut. The Rheon company of

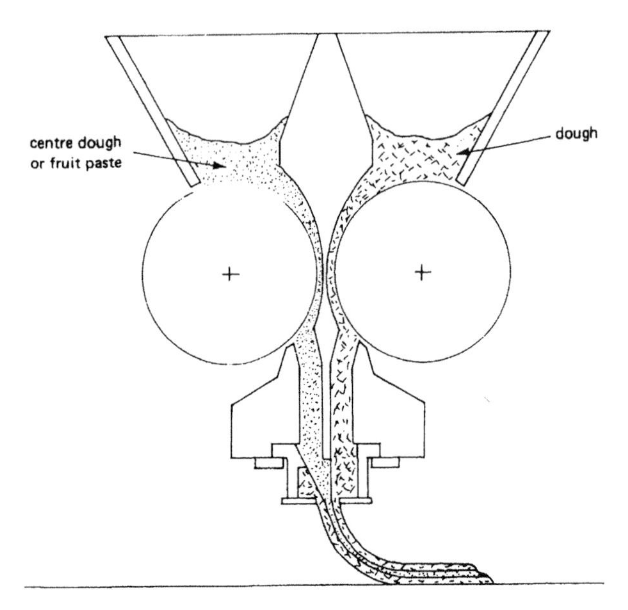

35 General arrangement of filled bar forming machine.

Japan has produced some innovative machines capable of co-extrusion with cutting systems that seal or encapsulate the filling within the outer dough. The cutting and sealing action is achieved either by specially shaped rotating discs or by iris cutters made of a plastic material. The machines are capable of handling a wide range of materials such as jam, minced meat, or very sticky dough but the outer dough must always be fairly soft and very short in texture.

8.4 Formation of dough and batter pieces by depositing

If the dough consistency is soft, smooth and almost pourable it is possible to produce deposited rather than wire cut dough pieces. The die plate is replaced with a set of piping nozzles. These nozzles are cone shaped and may have patterned ends to give strong relief to the extruded dough. Individual deposits are achieved by raising and lowering the oven band to coincide with intermittent extrusion which is achieved by intermittent movement of the feed rolls under the hopper. When the band is raised to be close to the nozzles, dough is forced out which spreads on the oven band. At the end of the extrusion the band is allowed to fall back and the deposit breaks away from the nozzle. The break away may be encouraged by reversing the action of the feed rollers producing a suck back at the nozzles. It is possible to produce fingers, swirls, circles and other shapes of dough pieces by using variable times and mechanisms which twist the nozzles. By synchronising a second depositor with the first, it is possible to deposit jam (or jelly) or another dough on or within the deposit made by the first.

There are a number of biscuits that are made from foamed egg batter. Deposits of this batter are made directly onto the oven band from a pipe over the band. The batter is pumped through the pipe and is extruded through small die holes which are opened and closed at intervals.

8.5 Dough piece weight control from extrusion machines

Wire cut and rout press extrusion machines are relatively simple in design. The pressure of the dough in the pressure chamber behind the dies is achieved and maintained by the friction of the dough on the feed rollers. This frictional force will change if the dough consistency (both in terms of softness and stickiness) changes. The general control of the rate of extrusion and therefore the dough piece weight is achieved by altering the speed of the feed rollers which then cause an increase or reduction in the pressure of the dough in the chamber beneath them.

There will be drag forces on the dough at the sides of the machine. It is therefore common to find that the pressure across the pressure chamber is not uniform and more dough tends to be extruded from the dies at the centre of the machine than from those at the sides. It would be very difficult to compensate for this by making dies with different aperture sizes. However, to effect some control of the rate of extrusion, and therefore the dough piece weight, adjustable restrictors are usually provided. These consist of plugs that can be screwed into the side of the die hole. As they are screwed in the size of the die reduces. Balancing the dough piece weights across the band is a complicated and often frustrating task. When the machine is started the pressure builds up in the chamber to a steady state. When the machine is stopped this pressure slowly falls by continued extrusion through the dies. Thus the effects of adjustments to the die restrictors can only be checked when the machine has been run for a few minutes. (It is usually not possible or safe to make restrictor adjustments while the machine is running.) Furthermore, if the consistency of the dough varies, either within a batch due to dough age or between batches, the variation in extrusion from the dies may be affected!

Dough piece weight uniformity is thus not a strong point with extrusion machines. The most useful engineering design that has been introduced to improve weight control is the provision of vertical divider plates within the pressure chamber. These prevent lateral flow of dough and more or less eliminate the hopper side drag effects. Ideally there should be a divider plate related to each or each pair of die holes. As the machine is designed to accept die plates with any number and size of die holes this means that the divider assembly should be changed to suit each die plate.

The size of batter deposits is controlled simply by the length of time that the dies are held open. There is often a variation of deposit weight across the band but this is generally less of a problem than for dough extrusion machines.

8.6 Starting and shut down of extruding and depositing machines

8.6.1 Start up

Before loading the machine with dough:

- Check that the machine is clean.
- Check that the correct die plate is fitted and is secure.
- Check that the wire is fitted correctly and is not broken.
- Start the machine and observe the action of the wire, does it pass closely below the surface of the die plate, adjust if necessary.
- Set the machine speed in accordance with the process audit chart for the biscuit to be made.
- Stop the machine.

Load dough into the hopper:

- Check oven is ready to receive product.
- Start the machine at the time set by the production schedule.
- Observe the extrusion action and the way in which the dough is falling, adjust speed if necessary.

Particularly look for differences between lanes of dough:

- Inform baker that production has started.
- When the machine has settled collect a set of dough pieces (length

of dough for rout press) from right across the band and measure the weight.
- Wait for confirmation from the baker that the biscuit weights are satisfactory, make adjustment to machine speed as required.
- Collect dough pieces from each lane and record the weight.
- Consider adjustment of the dies if significant weight variations are observed between lanes.
- Inform baker as and when any adjustments are made to the dough piece weights.

8.6.2 Shut down

- Run out the dough from the hopper.
- Stop the machine.
- Inform baker as last dough pieces enter the oven.
- Remove the die plate for cleaning.
- Clean the hopper and feed rolls.

8.7 Troubleshooting

8.7.1 Wire cutting

- Dough pieces are not falling uniformly and often touch another piece.
  - this is a common problem and it may not be possible to cure it completely. Adjust the height of the die plate over the band to see if an improvement can be made. Check that the cutting wire is tight and that it is passing close to the face of the die and is completely cutting through the dough extrusion before dropping down for its return stroke
  - consider fixing a buffer plate just beyond the stroke of the wire so that the cut pieces that are pushed back by the wire hit this and fall in a more controlled way
  - try adjusting the dough consistency
- Dough pieces are inverting after cutting.
  - this problem is related to the previous one. There is no problem if all the pieces are inverting uniformly as there is no correct way up for the cut pieces! Try adjusting the dough consistency if a change is needed
- Dough pieces are too thick/heavy in the centre relative to the edges.

  - the very nature of extrusion will give pieces thicker in the centres. This effect will be more pronounced with large dies and with softer doughs. Try adjusting the dough consistency
  - it is possible to design dies with a centre piece which restricts extrusion in the centre. There are some very large extrusions that are made in a ring. The dough flows and fuses into a more or less flat and complete biscuit before and during baking

- There are periods when pieces at one side are much smaller than they should be.
  - if the occurrence is periodic and irregular suspect the dough feeding arrangements. Is the level in the hopper running low or is the dough bridging? If it is the latter case try keeping the level of the dough in the hopper uniformly lower.

8.7.2 Rout pressing and co-extruding

- The edges of the extrusions are ragged and not smooth.
  - the dough is too dry. If the occurrence is intermittent it could be the age of the dough is the cause. Make the dough a little softer and stand it for at least 30 min before it is used. In this way the difference between the start and end of the batch of dough will be less pronounced in terms of consistency
  - the dough is too tough. It has been overworked either in the mixer or in subsequent handling. Consider reducing the final mix time and ensuring that the dough stands before transfer to the extruding machine or any other dough feeding machinery

- The size of the extrusion ribbons vary with time.
  - this could be due to variations in dough consistency either between batches or within a batch due to dough age
  - is the level of dough in the feed hoppers being maintained?
  - is the level of scrap being added at the mixer constant?

- Setting the ratio of centre fill to the outer dough is a problem.
  - it is normally not possible to measure the feed rate of the centre fill material in isolation. The feed rate must be inferred by difference. First start the feed of the outer dough, let the machine settle and sample to determine the dough weight per time or length. Now start the centre fill, allow the machine to

settle and repeat the sampling and weighing technique. Repeat the procedure a few times as small variations in the two feed rates can be expected
- check the height of the nozzle relative to the oven band.

8.7.3 Depositing

- The deposits will not break away from the nozzle at the end of the extrusion period.
  - the dough is either too soft or too tough. Attend to the water level in the mixing and ensure that the final mixing time is minimal. Watch also for handling procedures after mixing that could be causing the dough to toughen.
- The dough pieces do not maintain their surface relief.
  - if this effect is being noticed before the dough pieces go into the oven the dough is probably too soft
  - if this effect is not noticed before baking but the biscuits have flowed during baking the formulation should be adjusted to reduce the effect. Refer to Manual 4, *Baking and cooling of biscuits*.

9 Care, cleaning and maintenance of machinery

9.1 Care

The equipment in the dough piece forming department involves many moving parts and considerable interlinking of machinery drives and speeds. It is of great importance that no machine is started without a check being made that persons are standing clear.

The main aspects of careful behaviour are described in Section 3, *Hygiene and safety aspects*.

9.1.1 Consideration of others

- Dough spilled or dropped onto the floor should be cleaned up immediately because slippery floors can be a hazard to you and other workers.
- If you have to work at a high level over a plant, take particular care not to drop or spill anything on someone who may be standing or working below you.
- If you see a fellow worker in a dangerous situation give a warning. If someone is in trouble or struggling to handle something give a hand.

9.1.2 Consideration of equipment

- There are normally many fabric conveyors used to transport dough and dough sheets, these wear, break and can fray at the edges. Early attention to worn and damaged conveyors could prevent an unscheduled stoppage of the plant with the expense and inefficiencies that would be involved.
- Cutters and moulding rolls are expensive pieces of equipment that require removal and handling. They can easily be damaged if the correct procedures are not observed. Do not attempt to move these if you do not have the correct equipment or are short of the

necessary assistance. Cutters and moulding rolls should never be placed on the floor.

- Make sure that all hand tools such as knives and scrapers are kept in safe places where they cannot fall into dough tubs or on dough conveyors.
- Process control instruments are very sensitive and are easily damaged by mishandling. Do not hammer or kick them! If difficulties or malfunctioning are encountered attend to the problem thoughtfully or seek assistance. If the malfunctions occur repeatedly report them so that engineering attention can be given before something breaks down completely.

9.1.3 Consideration of possible contamination of the dough

- Make sure that floor dirt or any other extraneous material does not fall onto dough or conveyors that carry dough. If you see a maintenance worker, for example an engineer or building worker working near to or above a plant see that protective covers or screens are set up to reduce the risk of extraneous material falling onto dough or dough carrying surfaces.
- The metal detector, sited before a rotary moulder, should be provided with a distinctive tub to catch any dough that it rejects. Ensure that any rejected dough is disposed of safely and is not returned as normal scrap dough.
- Insect and rodent infestation is a particularly serious source of contamination especially in the dough forming department. Precautions should be in place to reduce the chance of insects, birds and animals entering a food factory but if they are seen report the matter to those responsible for controlling pests and cleaning.

9.2 Cleaning and maintenance of dough forming machinery

- Cleaning of dough forming machinery is the responsibility of bakery operators. Some aspects of maintenance are also involved. There are particular cleaning tasks that must be done at the end of each production run. Make sure that they are done as delay will make the job more difficult when the dough has dried and could hinder a smooth start to the next production run.
- Pay particular attention to the cleaning of sheeters, particularly the grooves on the forcing rolls. This is not an easy task but dough

which dries and hardens can cause a lot of problems and perhaps contaminate the next production run.

- The process of cleaning a machine will also provide opportunities to look for repairs that should be done or reported.
- The web tensions should be relaxed and dough which has collected on the drive and support rollers should be scraped off.
- Check that there is no build up of dough on the rubber roll(s) which act as anvils for the rotary cutters and if these need cleaning do so with care so that the rubber is not damaged.
- At the end of a run always relax the pressure between the cutting rolls and the anvil(s).
- All webs should have scrapers on their return runs and these should be checked and catch trays emptied.
- The webs themselves should be examined and faulty joints repaired as necessary. Due to wear at the edges the webs often become ragged and loose threads appear. Trimming with scissors or a sharp knife prevents these threads finding their way into the dough. Badly worn webs will, of course, have to be replaced with new ones.
- It is particularly necessary to remove and clean thoroughly the cutters, whether of rotary or reciprocating type. The cutters are heavy and expensive so suitable lifting equipment is needed to remove them from the plant and specially constructed racks or trolleys should be available to take them away or to store them. Cutters should never be placed on the floor.
- Rotary cutters can usually be cleaned adequately with stiff nylon brushes and compressed air blast, but the reciprocating cutter blocks and moulding rolls should be soaked in warm water and then blasted clean with steam. It is important to check that all dough residues are removed from behind the ejector plates of reciprocating cutters.
- Dough left around the machines or within them attracts vermin and insects and these, or their droppings, may later contaminate dough. Good supervision will insist that dough dropped under machines is cleared away regularly, but especially at the end of a production run.
- Look for leaks of water, grease or dust. Precisely report and record the problems.
- Your vigilance could ease your the job and that of your colleagues. It could also save much expense, inefficiency and frustration.

Useful reading and additional study

Almond N (1989) *Biscuits, Cookies and Crackers, Volume. 2 The biscuit making process*, Elsevier Science, London.

CABATEC (1988) *An Introduction to Biscuits*, An audio visual open learning module Ref. S8, The Biscuit, Cake, Chocolate and Confectionery Alliance, London.

Manley D J R (1991) *Technology of Biscuits, Crackers and Cookies*, 2nd edition, Woodhead Publishing, Cambridge.

Index

Printed and bound by CPI Group (UK) Ltd, Croydon, CR0 4YY

08/05/2025

01864833-0001